Food Industry Design, Technology and Innovation

The *IFT Press* series reflects the mission of the Institute of Food Technologists—to advance the science of food contributing to healthier people everywhere. Developed in partnership with Wiley Blackwell, *IFT Press* books serve as leading-edge handbooks for industrial application and reference and as essential texts for academic programs. Crafted through rigorous peer review and meticulous research, *IFT Press* publications represent the latest, most significant resources available to food scientists and related agriculture professionals worldwide. Founded in 1939, the Institute of Food Technologists is a nonprofit scientific society with 18,000 individual members working in food science, food technology, and related professions in industry, academia, and government. IFT serves as a conduit for multidisciplinary science thought leadership, championing the use of sound science across the food value chain through knowledge sharing, education, and advocacy.

IFT Press Advisory Group

WILEY Blackwell

Food Industry Design, Technology and Innovation

Helmut Traitler

Birgit Coleman
Connections Explorer, Swissnex San Fransisco

Karen Hofmann
Chair of Product Design, Director of the Color, Materials, and Trends Exploration Laboratory

Published by John Wiley & Sons, Inc., Hoboken, New Jersey
Published simultaneously in Canada

The contents of this work are intended to further general scientific research, understanding, and discussion only and are not intended and should not be relied upon as recommending or promoting a specific method, diagnosis, or treatment by health science practitioners for any particular patient. The publisher and the author make no representations or warranties with respect to the accuracy or completeness of the contents of this work and specifically disclaim all warranties, including without limitation any implied warranties of fitness for a particular purpose. In view of ongoing research, equipment modifications, changes in governmental regulations, and the constant flow of information relating to the use of medicines, equipment, and devices, the reader is urged to review and evaluate the information provided in the package insert or instructions for each medicine, equipment, or device for, among other things, any changes in the instructions or indication of usage and for added warnings and precautions. Readers should consult with a specialist where appropriate. The fact that an organization or Website is referred to in this work as a citation and/or a potential source of further information does not mean that the author or the publisher endorses the information the organization or Website may provide or recommendations it may make. Further, readers should be aware that Internet Websites listed in this work may have changed or disappeared between when this work was written and when it is read. No warranty may be created or extended by any promotional statements for this work. Neither the publisher nor the author shall be liable for any damages arising herefrom.

For general information on our other products and services or for technical support, please contact our Customer Care Department within the United States at (800) 762-2974, outside the United States at (317) 572-3993 or fax (317) 572-4002.

Wiley also publishes its books in a variety of electronic formats. Some content that appears in print may not be available in electronic formats. For more information about Wiley products, visit our web site at www.wiley.com.

Library of Congress Cataloging-in-Publication Data

Traitler, Helmut.
  Food industry design, technology and innovation / Helmut Traitler, Birgit Coleman, and Karen Hofmann.
    pages cm—(Institute of Food Technologists series)
  Includes bibliographical references and index.
  ISBN 978-1-118-73326-4 (hardback)
  1. Food—Packaging. 2. Food presentation. 3. Branding (Marketing) I. Coleman, Birgit.
  II. Hofmann, Karen. III. Title.
  TP374.T73 2014
  664′.09—dc23   2014025845

Cover images: Bulb icon © grmarc /iStockphoto;
          Strawberry icon © robuart /iStockphoto.

Printed and bound in Singapore by C.O.S. Printers Pte Ltd

10 9 8 7 6 5 4 3 2 1

Titles in the *IFT Press* series

- *Accelerating New Food Product Design and Development* (Jacqueline H. Beckley, Elizabeth J. Topp, M. Michele Foley, J.C. Huang, and Witoon Prinyawiwatkul)
- *Advances in Dairy Ingredients* (Geoffrey W. Smithers and Mary Ann Augustin)
- *Bioactive Compounds from Marine Foods: Plant and Animal Sources* (Blanca Hernández-Ledesma and Miguel Herrero)
- *Bioactive Proteins and Peptides as Functional Foods and Nutraceuticals* (Yoshinori Mine, Eunice Li-Chan, and Bo Jiang)
- *Biofilms in the Food Environment* (Hans P. Blaschek, Hua H. Wang, and Meredith E. Agle)
- *Calorimetry in Food Processing: Analysis and Design of Food Systems* (Gönül Kaletunç)
- *Coffee: Emerging Health Effects and Disease Prevention* (YiFang Chu)
- *Food Carbohydrate Chemistry* (Ronald E. Wrolstad)
- *Food Industry Design, Technology and Innovation* (Helmut Traitler, Birgit Coleman, and Karen Hofmann)
- *Food Ingredients for the Global Market* (Yao-Wen Huang and Claire L. Kruger)
- *Food Irradiation Research and Technology, second edition* (Christoper H. Sommers and Xuetong Fan) *Food-borne Pathogens in the Food Processing Environment: Sources, Detection and Control* (Sadhana Ravishankar, Vijay K. Juneja, and Divya Jaroni)
- *Food Oligosaccharides: Production, Analysis and Bioactivity* (F. Javier Moreno and Maria Luz Sanz)
- *Food Texture Design and Optimization* (Yadunandan Dar and Joseph Light)
- *High Pressure Processing of Foods* (Christopher J. Doona and Florence E. Feeherry)
- *Hydrocolloids in Food Processing* (Thomas R. Laaman)
- *Improving Import Food Safety* (Wayne C. Ellefson, Lorna Zach, and Darryl Sullivan)
- *Innovative Food Processing Technologies: Advances in Multiphysics Simulation* (Kai Knoerzer, Pablo Juliano, Peter Roupas, and Cornelis Versteeg)
- *Mathematical and Statistical Methods in Food Science and Technology* (Daniel Granato and Gastón Ares)
- *Microbial Safety of Fresh Produce* (Xuetong Fan, Brendan A. Niemira, Christopher J. Doona, Florence E. Feeherry, and Robert B. Gravani)
- *Microbiology and Technology of Fermented Foods* (Robert W. Hutkins)
- *Multiphysics Simulation of Emerging Food Processing Technologies* (Kai Knoerzer, Pablo Juliano, Peter Roupas and Cornelis Versteeg)
- *Multivariate and Probabilistic Analyses of Sensory Science Problems* (Jean - François Meullenet, Rui Xiong, and Christopher J. Findlay)
- *Nanoscience and Nanotechnology in Food Systems* (Hongda Chen)
- *Natural Food Flavors and Colorants* (Mathew Attokaran)
- *Nondestructive Testing of Food Quality* (Joseph Irudayaraj and Christoph Reh)
- *Nondigestible Carbohydrates and Digestive Health* (Teresa M. Paeschke and William R. Aimutis)
- *Nonthermal Processing Technologies for Food* (Howard Q. Zhang, Gustavo V. Barbosa-Cánovas, V.M. Balasubramaniam, C. Patrick Dunne, Daniel F. Farkas, and James T.C. Yuan)
- *Nutraceuticals, Glycemic Health and Type 2 Diabetes* (Vijai K. Pasupuleti and James W. Anderson)
- *Organic Meat Production and Processing* (Steven C. Ricke, Ellen J. Van Loo, Michael G. Johnson, and Corliss A. O'Bryan)
- *Packaging for Nonthermal Processing of Food* (Jung H. Han)
- *Practical Ethics for the Food Professional: Ethics in Research, Education and the Workplace* (J. Peter Clark and Christopher Ritson)
- *Preharvest and Postharvest Food Safety: Contemporary Issues and Future Directions* (Ross C. Beier, Suresh D. Pillai, and Timothy D. Phillips, Editors; Richard L. Ziprin, Associate Editor)
- *Processing and Nutrition of Fats and Oils* (Ernesto M. Hernandez and Afaf Kamal-Eldin)
- *Processing Organic Foods for the Global Market* (Gwendolyn V. Wyard, Anne Plotto, Jessica Walden, and Kathryn Schuett)
- *Regulation of Functional Foods and Nutraceuticals: A Global Perspective* (Clare M. Hasler)
- *Resistant Starch: Sources, Applications and Health Benefits* (Yong-Cheng Shi and Clodualdo Maningat)
- *Sensory and Consumer Research in Food Product Design and Development* (Howard R. Moskowitz, Jacqueline H. Beckley, and Anna V.A. Resurreccion)
- *Sustainability in the Food Industry* (Cheryl J. Baldwin)
- *Thermal Processing of Foods: Control and Automation* (K.P. Sandeep)
- *Trait-Modified Oils in Foods* (Frank T. Orthoefer and Gary R. List)
- *Water Activity in Foods: Fundamentals and Applications* (Gustavo V. Barbosa-Cánovas, Anthony J. Fontana Jr., Shelly J. Schmidt, and Theodore P. Labuza)
- *Whey Processing, Functionality and Health Benefits* (Charles I. Onwulata and Peter J. Huth)

Contents

Author Biographies

Helmut Traitler has a PhD in Organic Chemistry from the University of Vienna, Austria. He was an Assistant Professor and Group Leader of a Research Team for Westvaco in Charleston, SC, USA, working in Vienna, Austria. He joined Nestlé Research in 1981 and later became a member of the Editorial Board of *JAOCS* (*Journal of the American Oil Chemistry Society*). At Nestlé, his roles have included Head of the Department of Food Technology, Head of the Combined Science and Technology Department, Head of Nestlé Global Confectionery Research and Development, York, UK, Director of Nestlé USA Corporate Packaging in Glendale, CA, Head of Nestlé Global Packaging and Design, Nestec Ltd., in Vevey, and V.P. of Innovation Partnerships at Nestec Ltd., working in Glendale, CA as well as Vevey, Switzerland. In August 2010 he co-founded Life2Years, Inc., a start-up company in the area of healthy beverages for the 50+.

Helmut is the Senior Innovation Connector for Swissnex San Francisco, a public—private partnership organization with offices in Singapore, Beijing, Bangalore, Rio de Janeiro, Cambridge, MA, and San Francisco, sponsored by the Swiss government. He is actively involved in technology spin-offs of mission-non-critical know-how for the Jet Propulsion Laboratory (JPL) in Pasadena.

He has most recently been involved in co-developing food products in the area of sports and is the author of more than 60, mostly peer-reviewed, scientific publications, and 25 international patents.

Birgit Coleman is a strategic thinker and Connections Explorer in her current role at Swissnex San Francisco. Her expertise includes recipes for growth through internal innovation and external strategic partnerships with the goal of building a disruptive innovation pipeline for the clients of Swissnex San Francisco. Prior to Swissnex San Francisco, Birgit worked for the energy drink company Redbull North America, and IBM in Vienna, Austria—her home country. She holds a Masters Degree in Business from the University of Vienna.

Karen Hofmann is Chair of the Product Design Department at Art Center College of Design in Pasadena, California, where she is instrumental in developing an innovative curriculum responding to the expanding role of design. Along with her colleagues she is responsible for identifying and shaping vision-casting projects with numerous corporate sponsors. She leads the innovative study abroad program with INSEAD international business school where designers collaborate with MBA students to develop new ideas for products and services. Karen has developed innovative educational models such as the DesignStorm™, a compliment to Art Center's traditional transdisciplinary studios, partnering with sponsors to explore future market opportunities. She also serves as the Director of the Color, Materials, and Trends Exploration Laboratory (CMTEL) responsible for creating the strategy and developing this resource at Art Center to develop a unique educational program around color, materials, and trends.

Forewords

When my former Nestlé colleague and one of the authors of this book, Helmut Traitler, had asked me to read *Food Industry Design, Technology and Innovation*, I was hesitant. Every year I receive dozens of unsolicited management books—with catchy titles and bold theses—and after skimming a few pages they wind up in a cupboard behind my desk, unread. But this one is different. The book's premise that the role of design is under-appreciated and marginalized in industry, and in particular the food industry (in which I have worked for over 30 years), hits home from page one. But what makes this text unique is its ability to analyze and synthesize complex concepts, and spark thoughts and actions through real-world examples. Design is clearly one of the most important factors of our lives: professionally and personally. *Food Industry Design, Technology and Innovation* may not have a catchy title, but the content is thought provoking and practical. This book does not sit on my cupboard with the others, but near my desk, with its pages dog-eared and notes written in the margins. Enjoy and design!

Chris Johnson
Executive Vice President
Nestlé SA,
Vevey, Switzerland
November 2013

FOOD DESIGN, AN INTRODUCTION

Eating and drinking are amongst the most precious of the experiences that make up our lives—both the primal sensory experience, and the social pleasure of breaking bread with friends and family. By designing the food and drink we consume, we have the potential to make these experiences even richer. But food design is complex. Planning, buying, transporting, storing, preparing, serving, eating, sharing, savoring, cleaning up, digesting, metabolizing, and incorporating into our bodies for our immediate energy level and for our long-term health are just some of the elements of the experience of eating and drinking. Further removed, the land, energy, and labor used in growing, transporting, processing, and selling food also have enormous impact on our lives and on the planet.

Food has always been designed, but generally not consciously, deliberately, and holistically. We appreciate the presentation and freshness at a fine local restaurant, we might notice the packaging on a supermarket shelf, and against our better judgment we are seduced by the taste of molecules concocted in a candy bar, but we don't connect them as a continuum of design. Adding to the complexity are the contradictions in our expectations about food. The food industry is global, but eating is local: we love to try local flavors when we travel, but at home we mostly eat according to our cultural norms. In the West we expect that food should be always available, and we bemoan our tendency to overeat. We shop for the lowest price and complain about industrial agriculture taking over the countryside. We are misled by the contradictions between the appearance of health and sustainability, and the actual impact of different food choices. The complexity of how we choose what to eat, and the opacity in the effects of those choices make it difficult for consumers to know what they want.

Over time, like any complex system, it will evolve to meet our emerging preferences. Design offers the possibility of accelerating the process for the good of all, especially the companies that lead the way. Designers and innovators have the unique ability to see into people's lives and to envision ways to improve them. Beyond just packaging, design is now reaching into the complex web that makes up our food supply and beginning to deliberately make it better. The ways of thinking about design have broad application and are becoming central to how companies compete. To succeed, food designers need to understand people and envision what they will want, they need to understand technology and systems and be able to show how to deliver what they have envisioned, and they need to understand organizations in order to make innovation happen in a corporation.

Helmut Traitler, Birgit Coleman, and Karen Hofmann's ground-breaking book is the future of food design. By integrating design, technology, and organization they offer a guide to the conscious, deliberate, and holistic design and innovation of food.

Harry West
Senior Partner, Prophet, New York City
Board Member of Continuum Design, former CEO,
Cambridge MA
January 2014

Acknowledgements

Many people have inspired me to embark on the journey to write this book and quite a few have helped me to bring it all together. Even if the list below is by far incomplete, I want to thank those most critical in this endeavor in a few words: Sam Saguy for his constant challenge and inspiration, Heribert Watzke for always having had yet another great idea, Philippe Roulet for his deep insights into packaging related topics, Kai Taubert for his guidance in open innovation topics, Nick Traitler for his tremendous support in rendering images and figures good to look at and understandable, with much patience and professionalism, Paul Polman for having strengthened my belief in innovation, Spencer Mackay for having taught me the value of design, the late Glenn Tsuiuki and more recently Martin Barmatz who have encouraged me to look beyond food into space, Thomas Beck who has given me the opportunity to teach and encourage young scientists and engineers (and to instill the drive for innovation), Susanne Lauber Fürst who always forced me to look for new ways, Thèrèse Meyer Traitler for having supported me throughout the entire project, and the publishers, Wiley and IFT for having believed in this book. Finally, I would like to thank my two co-authors Birgit and Karen for the trust they had in this project and their great contributions, as well as Pierpaolo Pugnale, aka pecub, for having put a big smile on our faces.

Part 1

The role of design and technology in the food industry

1 Design and technology

Everything changes and nothing remains still.

Plato, from The Dialogues, Cratylus, paragraph 402, Section a.

ALL IS FLOWING: πάντα ρεῖ

This is not a book about design, it is not a design text book, but rather a collection of thoughts, analyses, and examples, paired with critical discussion as to how design, combined with innovation and smart technology can be put to good use in the consumer goods industry, especially in the food industry.

The food industry is probably not the first port of call when you look for good design, let alone great design.

Much comes down to the basics, i.e. manufacturing, marketing, distributing, and selling industrial food products in the most efficient and, for consumers, affordable ways, food that tastes good, is safe and nutritious, and is inexpensive. Most often, this does not allow for design in any form, with exceptions such as graphics and industrial design when it comes to packaging, or media design where communications and marketing are concerned, although the latter is even more rare than the former.

To say it right from the start, we feel that design is grossly and almost "criminally" underused in the food industry. If it is used, it is mostly for the more trivial matters such as logos, fonts, and colors. This is not to say that these are unimportant or easy tasks, but they are, by far, not enough. If designers are present in the activities of the food industry, they are rather kept at arm's length and in check by the marketing folks.

This book intends to make the point that not only are designers of many more disciplines tremendously important for the present and future successes of food corporations, but that designers should actually play an active and decisive role at the executive board level of any food company that strives for greater heights and greater success. It is the thought process of designers that is perfectly complementary to the other, more traditional thought processes already represented at board level, the MBAs, the finance experts, the marketing experts, the odd technical expert, and the lawyers. Fresh blood is dearly needed and designers can supply it!

Food Industry Design, Technology and Innovation, First Edition.
Helmut Traitler, Birgit Coleman and Karen Hofmann.
© 2015 John Wiley & Sons, Inc. Published 2015 by John Wiley & Sons, Inc.

The title of this chapter states that "all is flowing," the ancient Greek πάντα ῥεῖ (panta rei) as Heraklitos of Ephesos formulated it around 500 BC and on which Plato elaborated in one of his "Dialogues." He was referring to water and life in general, which is pretty large and daunting in itself. What it refers to here is the fact that over the years of development in the manufacturing and especially consumer goods industry, the notion of design has always been present and has always changed in importance and relevance, very much depending on the importance of form and function in the context of manufactured goods.

When we look at the early days of the automotive industry, over 100 years ago, function and functionality, the surmounting of technical hurdles, and the solution of technical challenges were the most important, overriding principles. Design was not at the forefront of the thought process of mechanical engineers or engineers in general, for that matter. Getting things done and making them work was more important. The technical challenges were large and difficult enough, so there was not much room for anything else.

With the advent of the production line, as introduced by Henry Ford to manufacture the famous and notorious Ford T models, design elements were becoming more important, and engineers would design shapes and forms that were very much based on functionality and the direct needs of the end user, the driver of the vehicle. We may call this approach "down-to-earth" design (DtE design), with little or nothing that goes beyond the straightforward desire to deliver something that works, is functional, and, most of all, is affordable.

The element of "it works" is fulfilled by the increasing competence of the engineers, the "functional" is the first sign of good design, and the "affordable" can clearly be linked to the production methods that were introduced for the first time in those days.

The next quantum leap in the history of design came with the invention of plastic materials and their industrial usage. The year 1907 marked the invention of "Bakelite," followed by the creation of "Formica" in 1913. However, it took more than 30 years before the synthetic materials, plastics, came into the mainstream and were heavily used in many different industries, most importantly in furniture and consumer goods (Lewin, 1991).

Very rapidly, the automotive industry developed a very strong and years-long love affair with increasing design elements of a non-functional nature, and this trend can very easily be seen in the American cars of the 1950s and 1960s, which really seemed to be the playground of designers, where they could realize their most extraordinary fantasies and dreams. Functionality was second and design elements even became dangerous additions to otherwise very functional vehicles.

When the Bayer company of Leverkusen in Germany invented polycarbonate in the late 1950s and early 1960s they also wanted to show its versatility beyond simple household goods such as glasses, plates, cutlery, bottles, baby bottles, and the like, which resisted far higher temperatures than other plastic materials at the time. The first-generation polycarbonate could be used at temperatures of up to almost 130 °C and was therefore quite unique in those days. But Bayer had another ambition too: it wanted to prove that its innovative plastic material could be used in the automotive industry, and not just in a few parts here and there, replacing steel and other metals, but in the entire car, from the engine bay, to the trunk, to the passenger cabin, and the body.

Thus, back in the mid-sixties, Bayer revealed the first, fully plastic (except, of course, the engine, exhaust, and gearbox) concept car. Long before its time, this concept car was a strong harbinger of things to come and led the way to an increased use of plastic materials in the car industry and automotive design.

If we follow this line, we can also observe that metal shaping knowhow became increasingly sophisticated, and plastic molding and steel or aluminum forming became concurrently and widely used technologies during these years and into present times, thus enabling designers to play with all these materials in many new and exciting ways.

As we said above, "all is flowing," and so is design or rather the approach to design and its perception and role in today's world. When you ask people, "What is design?", most link it to fashion. It's the standard answer of the average, "uninitiated" person, which you will get probably two times out of three. What does this mean? Well, most likely, the world of fashion has made a better job in promoting one of their more important elements of success, which is, undoubtedly, design. This also means that other industries have not been equally successful.

Let us again use the automotive industry as an example. After the "sins of the 50s," the design of cars was very much toned down and safety elements became *the* driving force. There was and is nothing wrong with this, as it has certainly helped to make driving safer these days. There was one factor, perhaps, that weighed even more in increasing drivers' safety and that was the introduction of random checks by the police, which was by design, but has nothing to do with design.

Remember the Volvo bumpers? Huge promontories with lots of rubber, definitely advancing the safety features on vehicles, and also certainly helping the shareholders of the rubber industry. Every car, all of a sudden, seemed to have them, even the MGB of 1975. Design, once again, was secondary, the driving force was functionality in the constructive disguise of safety. Remember, all is flowing and so is design.

At the end of the day, it's all about form and function and the struggle to get a harmonious balance between these two basic and crucial elements. So, fashion has done a good job. Who else? Let us expand a little bit on education, on schools and colleges, places where design is taught and where the future (or present) designers will come from (came from).

Design schools probably existed longer than we would imagine and we may argue that Guarneri's or Stradivari's workshops with many students or apprentices were already design schools. What better item is there than a wonderful and well-sounding violin to show the fantastic equilibrium between form and function. And it goes farther: the choice of materials, not only the woods but also the paints, the mixtures of natural ingredients, the way of cutting, gluing, sanding, applying, and finishing off such an instrument. Students were taught and they learned how to do this work, how to create these objects of musicians' desires, and how to optimally balance form and function. If this is not design, and if this is not a school of design, then we would not recognize reality.

But we can go back even further: the Sumerians, Hammurabi's mask, and other related relics were not only works of devotion and art, but totally designed, the marble lions of Delos, the buildings in ancient Greece, where architecture and art, design ultimately, came together, joined forces, and left wonderful signs of design; all this was taught, and students, apprentices, and future masters went through schools that we would call design schools by today's standards.

All is flowing!

So, back to the present day, because that's what has to interest us in the context of this book, namely the role of design in today's industry, especially in today's food industry, and how this links to technology and innovation in general. And, back to the question, how today's designers are educated and where this takes place and what is the emphasis on the different topics taught in such schools. Let's show a few examples and discuss the various orientations of these institutions.

Design schools can be found almost all over the developed world. Not so long ago in 2004/2005, Lim Kok Wing opened his "College of Creative Technology" in Cyber Jaya not far away from Kuala Lumpur in Malaysia. The school very quickly drew almost 4000 students with more than two-thirds from South East Asia and the remainder from Asia Pacific (Australia, New Zealand).

The location is a very interesting one, being one corner of the triangle "Kuala Lumpur—Airport and Formula 1 circuit—Cyber Jaya/Putra Jaya", the last being the seat of the prime minister and the government of Malaysia.

Cyber Jaya is the location of many more university colleges, the Lim Kok Wing School being just one, yet a rather large school given that it is a design college. The school intrigues by its architecture as well as the diversity of offerings, of branches that are taught. It's in a tropical environment, therefore the building has a center space that is just like a big tent, wide open on two sides and a meeting place for students from all the different design branches. And there are numerous branches, the usual ones from graphics to industrial design, but also the less typical ones such as hair design (yes, hair design is design too) to fashion and media design.

Staff and students proudly showed us different communication campaigns that the media design students did in the past for well-known politicians and that were a great example of what design can be, namely a vehicle to shape not only an object but even more so a process, an evolution, something that is on its way to being shaped and created.

It is not so much the final object and its perfect shape, form and function that are of importance, but almost more importantly the way, directional as well as procedural, that one gets there.

This aspect of "how we get there," of how "all is flowing" is a very crucial aspect of design and how it can apply to society, how it can apply to the individual, and, ultimately, how it can and will have to apply to corporations, especially those that operate in the fast-moving consumer goods area, such as the food industry.

As we are already, at least mentally, in South East Asia, let's turn to another example in the region, namely the LASALLE College of the Arts school in Singapore, one of two schools that will be briefly mentioned and discussed in this chapter. The LASALLE school, founded in 1984, very quickly became a school for drama and design, already having the creative vision of combining the arts, especially dramatic arts, with innovative design. When they moved into their newly constructed premises in early 2007, a very prominent place, almost a showcase, was given to dance; we could almost call it the "design of movement," basically the first thing a visitor can see when approaching the impressive modern and functional multi-tower building that houses the LASALLE school.

One can find all or most of the other traditional design branches, with industrial design being a very important leg. Staff members come from many different backgrounds, but given the British heritage, many of them come from the British Isles with educational backgrounds from Newcastle and other British design schools. Again, the intriguing part here is not only the diverse and modern approach to design, but the inclusion of other artistic activities, such as drama and dance, into the definition of what design is and can be, namely a truly inclusive approach.

Just a few miles away from LASALLE we find the Nanyang Polytechnic, a large engineering college with over 12,000 students, next to the NUS (National University of

Singapore), the largest university in Singapore. Nanyang has a small design department that is especially active in furniture design, but also packaging design, and has strong capabilities and competences in tooling, and mould design and creation for fast prototyping, which is an increasingly important activity in today's world of object design.

To finish off our short excursion to Singapore, it is certainly worthwhile to mention that there is a very attractive and active Red Dot Museum in Singapore that shows many examples of already proven and industrialized design objects, as well as some more exotic and visionary design solutions to recycling packaging materials, such as PET bottles for building purposes.

Another top-class design institution can be found in Pasadena, California, namely the Art Center College of Design (Art Center) with two campuses and several thousand students. In the past, the Art Center was mainly famous for its automotive design. Not so long ago, approximately 50% of all vehicles that were out on the streets were at one time designed by former students of the Art Center and there is still an annual show of such vehicles, depicting the strength and heritage of the Art Center in the world of automotive design. But that's not all, as we know by now, all is flowing, and so are the directions and orientations that even such a prestigious and well-established school as the Art Center have taken.

Being close to many different stimuli such as the arts (movies, entertainment, theme parks, etc.) as well as science (Caltech, JPL, UCLA, USC), the Art Center collaborates with all or most of the representatives of these very different areas. The Art Center has also added two new approaches to their traditional object design: the first one really touches upon the question of process versus object, and striking the right balance, whereas the second concerns the planning of new educational collaborations with business schools, especially with schools such as INSEAD in Versailles (France) and Singapore. Several courses were organized that combined the two worlds of business and design, with students from both branches.

There were even plans to enlarge this educational collaboration by including engineers from, for example, Caltech, but at this point in time it is unknown how far this really went. The idea is of course a wonderful and fascinating one; we know from experience and from historic examples that real creativity and innovation happens at the borderlines, the fault lines between different disciplines, and the theory goes that by increasing the number of fault lines we can dramatically increase the number of innovative events.

There is one last good example of a creative design school with lots of interfaces to scientific and business areas, namely the ECAL (Ecole cantonale d'art de Lausanne) in Switzerland. The school is located in a former industrial building that was transformed to fit the needs and requirements of a design school and has lots of space for work, as well as encounters. It is in the heart of what is called the "arc lémanique," which is the northern and center shores of Lake Geneva, especially around the city of Lausanne.

Lausanne sports two top-class universities the EPFL (École Polytechnique de Lausanne), as well as the UNIL (University of Lausanne). EPFL in global university rankings is one of the top 50 schools in the world and is amongst the 10 best in engineering.

All is flowing, and especially the old dividing lines between different, seemingly very far apart disciplines, such as design and science and engineering are starting to flow into each other, and are disappearing. ECAL has a strong graphics design department, but mainly lives on the strong interactions with the technical world, as well as the very numerous businesses in the "arc lemanique," which is home to a great number of large global companies. It is again the fault line between traditional design disciplines and other branches, such as business and technology that is a fertile playground in which to use the full potential of design and its capabilities to create, shape, form, and make functional in the most efficient, simple, and aesthetic ways.

Chapter 4 discusses and describes in much detail design in academia, the role that design education plays, and the influence that it has on society, but also on the corporate world.

HOW DESIGN INFLUENCES OUR LIVES: FORM AND FUNCTION

We have actually not yet talked about what design really means and can present.

Merriam—Webster's Collegiate Dictionary (10th edn 2001, p. 312) actually gives three definitions that are of relevance to "our" design definition:

1. The arrangement of elements or details in a product or work of art
2. A decorative pattern
3. The creative art of executing aesthetic or functional designs

These are all very good definitions and we are not here to argue with them. However, they mainly describe an action (1 and 3) or a situation (2). Yet, design is more than that. One could argue that everything is design or to be more specific, is designed. Just to be clear, there are no theological undertones here, strictly observationally speaking, nature shows us that form and functionality perfectly well go together, and are the way that nature is designed.

The harmonious balance between form and function is the most basic underlying principle as to how a plant, for instance, is constructed: a mixture of fibrous substructures, lignified parts to give strength and the possibility to grow higher, channels to transport water with nutrients, branches to spread out to enable leaves to find the most exposure to light, rain as well as oxygen and carbon dioxide, the latter two being most critical for the growth and survival of any plant.

Surface structures of leaves are such that a leaf can do two things: adsorb and absorb carbon dioxide to transmit it to other parts in the leaf for subsequent photosynthesis, i.e. transforming CO_2 and water with the assistance of light and chlorophyll within the chloroplasts of the leaf into carbon backbone molecules, namely sugars, necessary for the plant to feed on, and oxygen, given back to the atmosphere.

Like every living being on our planet, the tree needs oxygen to breath as well, which happens during dark periods, i.e. at night, and that equally happens through the surface and inside the cell structure of the leaf, so this is a two-way street, a structural element with functions to absorb and desorb molecules of vital importance. This is a perfect example where form and function play wonderfully together and there are many more, thousands, if not millions, such examples in nature with similar form—function elements.

We can therefore truthfully say that basically everything around us is designed in one way or another, and all this with probably only one principal design element, which is the best harmony between form and function. By looking again at the origin, nature and application of "man-made" design, we can easily say, by simple personal observation, that most design and designed objects do not follow this principle and yet, it would be so easy.

There are few examples of really well-designed objects that respect the form—function principle, but not many. We will look at these further below.

Let us first look at the long list of design that is clearly not taking this into account and that we could classify as design for design's sake: it's just for the eye, which per se is not a real drawback, as long as the designed feature makes sense one way or another, either for aesthetic or economic reasons, provided it does not go against the fundamental principles described and discussed above. Automotive design of the 1950s and 1960s can be classified under the headline "Trying to please a specific customer base." Lots of chrome, exorbitant shapes, so-called trendy colors, and similar attributes. Of course, one can argue that there is still a certain form—function principle to be seen, at least to some degree, but it is very well hidden under a large amount of heavy design overload.

There are many more such examples in more recent times, not only in the automotive field; the evolution of graphics design through the last couple of decades can also be used as an example. When looking at newspaper logos and typesets of the last 30 to 40 years we can clearly see a trend to a more sober, "cleaner" image, plus the introduction of color into mass-produced newspapers and other printed items. There are two main reasons for this. Firstly, technology has evolved and up to 11-head large industrial printers for multiple colors and other graphic elements are a reality today, although still fairly expensive to use to their full potential. Secondly, public taste has evolved and tends to be more sober and simple, despite the fact that we can find hundreds of different fonts in the word processors of our computers. It appears that more choice has led to a lower uptake! How come? Like every aspect of our lives, much of this is linked to fashions, fashions that come and go and come again, and so on. And we cannot even state, with certainty, that fashion is only driven by one side that has an interest in change so that new products with new designs can be created, manufactured, and sold. It is not as simple as that. Fashion is driven from many angles. It is certainly driven by the party that wants new economic success, but it is also driven from the consumers' end and probably many more "interested parties." It goes beyond the scope of this chapter and also this book to analyze all elements and factors that drive fashion and fashion changes, but it is important to understand that fashion is an extremely important element of design and ultimately design choices, often overriding our well-established form—function principle.

If fashion drives design then we should also deduce that design drives fashion; and this is certainly the case, very much so. As with the infamous chicken-and-egg, it is unclear what comes first and any speculation might be futile. However, we should look at the two sides, just for the argument's sake.

When Volvo introduced the well-known enormous safety bumpers, it subsequently drove a new fashion of enormous, rubber-lined bumpers on almost every car, even on an MGB, as mentioned earlier.

When the public expectation drove a fashion to smaller and more multifunctional objects, Apple created its first iPhone. It is arguable whether public expectations really came first, but it is a likely description of the situation and the argument can be made that Apple was better at reading consumers' desires, although they would probably not entirely agree: their argument would be that they were just more visionary and foresightful than anyone else. Whichever way, it just shows that the question whether "fashion drives design" or "design drives fashion" cannot be answered unequivocally. This is an important notion to keep in mind, now and for the following sections of this chapter, as well as the entire book. We will discuss this aspect as we discuss the important elements of this book, namely the influence of design and design thinking on the fast-moving consumer goods industry, especially the food industry.

If fashion or rather fashions and design are so interwoven, then we need to discuss in more detail the important interactions between design and the "receiving end," the consumers, the different communities and, why not, society at large. It is a very visible fact that fashion and design influences the look of a society: what clothes do we wear, what accessories do we use, what haircut do we sport, what gadgets do we surround ourselves with, what cars do we drive, what condominiums do we live in, what furniture do we sit or sleep on, what materials are used in our condos or houses and who tells us what is fashionable and therefore desirable?

THE HGTV EFFECT

HGTV or Home & Garden TV is a very successful, originally Canadian TV station. It features shows like House Hunters, House Hunters International, Income Property, Love it or List it, Property Brothers, and many more, all linked to homes, very few to gardens anymore. There is, however, one very strong common element to all these shows: they drive desires by driving fashions by insinuating, more than subliminally, that certain materials and styles are a must for the masses today. There are always four recurring messages and they come in all "colors and variations". First, it has to be open-plan living with lots of kitchen space because "we love to entertain"; most often, "entertainment" is then depicted by cutting a few fruits in a gigantic kitchen! Second, appliances *have* to be stainless steel and the cooktop *has* to be gas! Induction heating, increasingly also used by professional chefs is just beginning to find HGTV's blessing. Third, countertops *have* to be granite; it's just what you have to have. More recently, quartz countertops have become fashionable too, and they certainly start to show the next trend. Fourth, floors have to be hardwood, and large tiles or marble in bathrooms. The funny thing is that these four rules seem to apply to any type of house, in any location, any style, and any price range.

So, who drives what? In order to answer this, we may have to dig a little deeper into the story and look back a few years. Increasing affluence, especially after WWII and the 1960s and 1970s has, amongst many other phenomena, such as two cars per family, more exotic vacations etc., also led to increased sizes of condominiums or houses. As an example, most of the new condos built in Switzerland before 1980 were below 1000 sq ft, since then, 80% of condominiums in Switzerland have a size of more than 1000 sq ft.

It may be argued that, eventually, such bigger space also led to the idea or the desire to spend more time at home and render one's place more cozy and stylish. That, of course was the opportunity for related industries to propose new materials, and out went the old, cheap stuff, like linoleum, parquet, carpets or small-sized tiles, and in came hardwood floors and large-sized tiles, the latter also because technological advances were made in that industry allowing for tile sizes of 25 inches or, more recently, even larger. Out went the old-fashioned laminate or tiled countertops and in came granite, almost exclusively these days, without much questioning whether it really makes sense and fits in the environment. Out went the old-style wardrobes, and, because houses and condos became bigger, there was space to put walk-in closets, ideally one for her and one for him, hiding all the stuff that would either go into wardrobes or be lying around in a not so orderly fashion. The next item on the increased size list was the kitchen. Kitchens more recently have become really fashionable spaces.

Most importantly these days, they have to be open to the dining and/or family room so that the person or the persons who prepare(s) the meal can communicate, visually as well as orally, with the rest of the family or their guests. It is suspected that this trend did not start as a feminist off-shoot, but rather it was first created in the restaurant industry.

In the early 1990s, Disney created open kitchen restaurants in all their theme parks. All kitchens were totally open to the dining room and guests could see the professional chefs at work. There is no proof that Disney was the first in this trend, but it was a very strong instigator and promoter [D. Hannig, Disney Food & Restaurants, personal communication, 1994]. Many restaurants followed this trend and many popular food chains in the USA have designed their restaurants in this open-plan fashion since. It can be deduced that once the trend was well established in the popular restaurant world, it would not take long before it was followed in the homes of fashion- and trend-motivated individual home owners.

In the early 2000s a new trend, the "cocooning trend" could be observed. After a peak of out-of-home eating of around 50% in the USA and other developed countries at around the same time, and with the improved home quality, space, and coziness, it was maybe to be expected that people, especially young professional couples, would discover their own space again and stay home more. Staying home more also means having friends and family over to one's home more often, thereby using the increased space and more open nature of one's kitchen and living area in a much more efficient way.

Here's a controversial, yet believed to be true, personal observation and statement: most kitchens in Italian homes are small, if not to say tiny, and yet, meals that come out of Mama's kitchen are, most of the time, close to divine. So the statement could be: "The tinier the kitchen, the better the meal"; for the reader to contemplate and discuss, maybe over a meal with your friends in the open-plan, large environment of your home?

So, let's get back to the initial question as to who drives what? From the short analysis above, it can be said that the answer is not simple and clear cut: most likely, because and as a result of the trends described above, HGTV, and similar programs, drive fashions and desires, and the consumers, the real-estate clients, see these as a must, to be followed almost religiously. Just think of the alternatives: what about white or colored appliances, following the color scheme of the surrounding kitchen, what about induction stove-tops, what about polished concrete, colored or gray, what about, what about, what about … ? Moreover, brass fixtures, tiles with large grout lines and electric coil stoves are seen as a thing of the 70s (and they are, no doubt about that!), but the point is, some of these just may fit into my very personal color and design pattern.

Wait, there are more must-haves of our day: large master suites with "tons of light" through extra large windows seem to be a must too, although one typically sleeps in a bedroom. Yes, there are other things too that happen there, but none of these require "tons of light," but they rather require "tons of window treatments" to get the room dark enough to sleep. All this is in no way a rant against HGTV, but just shows a rather large number of really striking examples as to how fashion is driven and desires are created and steered, and the relevant industry has all the responses and solutions readily available, and available fast.

When watching other shows like "Grand Designs" and "Grand Designs Australia," you observe similar trend patterns, although they are subtly, often strikingly different from the ones constantly repeated by HGTV. Kevin McCloud, host and presenter of the show, makes a very fine job when insinuating certain trends, but ultimately the results are the same: fashions are created and desires are driven, not so much towards the "four HGTV commandments," but here we are subjected to the use of exotic building materials such as car tires, inside—outside houses, energy independence, all-transparent glass houses, and, as the title of the show says, grand designs, not just grand meaning big, but grand meaning daring and different, and demonstrating that grand yet humble architecture shapes the people who live in such places. It has to be said that what sticks most when watching shows like Grand Designs is the word sustainability, which, after all, is not such a bad key word to retain. However, the patterns, as said above, are similar, and people will follow these, more or less convinced that this is the way to go, the way to follow.

How much chance does the individual have to escape these fashion musts and how daring can and will the individual be in creating his or her own fashions and patterns in any situation of their lives, especially when it comes to design and design-related matters? It is probably the same as with all fashions, from clothes to shoes to cars to computers, smart phones, cameras, and many more items of our daily lives: we can hardly escape what is dictated by many players, especially the consumer goods industry and everything linked to it, such as advertising, media, and, increasingly, the social media.

We will discuss the role of social media in a later chapter, but we can say this much: social media are already replacing and will increasingly replace traditional communication channels, and it is debatable if the importance of traditional media and their influence on consumers will develop in parallel with the rise of social media channels. Today, the number of hits on YouTube or the number of followers on Twitter is an increasingly important measure for present and future successes of products or fashions, as well as new behavior patterns in many communities and societies.

It can safely be concluded that there is no escape from prevailing trends, and maybe this is not such a bad thing. There will always be individuals who can and will escape and create their own, new space, which may or may not become larger and more successful, and eventually replace existing trends to such a degree that they will take over the public at large and become mainstream. It is not bad at all and gives hope to the outsiders who will, as in the past, drive future innovation, together with the many other drivers in this landscape of fashion and innovation.

DESIGN IN THE FOOD INDUSTRY

As stated in the very first sentence of this book, this chapter, the food industry is maybe not the first port of call when it comes to looking for good design, or any design at all. This sounds like a strong statement, but from personal experience it can be said that it comes very close to reality, obviously with many good exceptions and examples that prove the contrary. Why has the food industry been so negligent in the design field? The answer is pretty simple: because it could.

But why could it? Well, because no one cared or at least didn't see this as an important, missing element. After all, food is just food and there are other fish to fry when it comes to food products.

Let us be more specific. Major criteria for industrial food products read like:

- Tasty and good, expected flavor and texture
- Safe
- Healthy and nutritious
- Convenient to prepare or finish

- Good value for money
- Little out-of-pocket spending and overall affordable (in many instances)
- And yes, there is packaging as well, and yes, hopefully it is not overpackaged
- Increasingly, sustainability and environmental responsibility, small carbon footprint are on this list as well.

It can be said that there are more criteria linked to good industrial food products, but the above is already a rather comprehensive list. It should also be stated that the individual importance of these criteria may differ from location to location, from culture to culture, and even the term "nutritious" may have different meanings, for instance for a country like Eritrea versus the UK, just as an example. Good nutrition in the UK may mean low fat with a good mix of fatty acids, or low sugar or salt, whereas in Eritrea it could just simply mean to get enough calories and salt, the latter to retain water in the body.

The one element in the list above that has the strongest and most logical connection is packaging and all packaging-related matters, although it should be said at the onset that good design does not stop at the packaging of a food product, but can and must apply to the food product itself. We will discuss several examples of this further below and in later chapters.

Let us first look in some detail into the role of packaging in the food industry and how this can relate to design, technology, and innovation. Traditionally, and from our own experience, packaging in the food industry was and maybe even still is, despite many good examples pointing to the contrary, the unloved child of the industry. When first being exposed to packaging in the food industry in 2002, and this after more than 20 years of having worked in parts of the industry that were concerned with new products, ingredients, and process development, the term "packaging" was not on the top 10 list of words being heard. A first encounter with packaging was accompanied by explanations such as: costs too much, can we reduce the amount of materials that we need?, it's a necessary nuisance, make it go away as much as possible. This pretty much explains that many in the food industry, and we specifically refer exclusively to the food industry, and not other branches of the consumer goods industry, saw packaging as a problem to make go away and not as the opportunity and defining element that it really can and must be.

In a very emotional and defining moment during a meeting with the president of the Japanese Association for the Elderly [Private Communication, Tokyo, November 2003], the role of packaging was discussed in great detail and particularly outspoken and important, if not essential, needs for packaging were listed and established.

Figure 1.1 depicts, in an overview, the various needs that were listed, but most importantly, good packaging was defined as being based on "universal" and "inclusive" design, meaning that such packaging takes note of the needs of all parts of the population, the very young and the very old, and by consequence, everyone in between. It has to be "friendly to human beings".

This is how we established principles of good packaging:

As can be seen in Figure 1.1, there are seven main elements, the criteria that are of primary importance when we speak about concepts of good packaging in the food industry. There is one increasingly important and even overriding criterion for good food packaging: it has to be environmentally friendly and respect the rules of sustainability. It goes without saying that many or most of these criteria apply to other members of the consumer goods industry as well, such as household goods, small appliances, audio and video media,

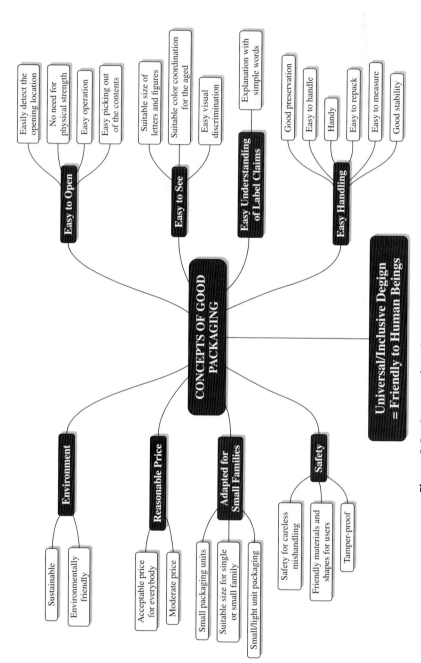

Figure 1.1 Concepts of good packaging in the food industry.

perfumes and fragrances, and similar areas. In discussing these criteria it should be noted that they are not listed by importance; they are all equally important in a general sense, and depending on the type of product, target consumers and geography might be weighted differently, but as a first approach, we do not weight these criteria differently.

Reasonable price

It is difficult to put an exact number on the value that a package represents for the consumer, but it is obvious that the more of the other boxes of good packaging are ticked, the better and more valuable the perception by the consumer. Anyhow, for the consumer it is difficult to judge what cost ratio of product/packaging any given packaged, industrial food product represents, and he or she, unenlightened, would have a hard time seeing that in a canned soda sold in the USA, the can costs more than its content, at least in most cases, and that in a bouillon cube sold in Nigeria, for instance, the content, i.e. the cube, outweighs the package cost wise. In the end, protection of the product is the major driving force, and that often has its price. In light of the above, the notions of "acceptable price for everybody" and "moderate price" are becoming more understandable and meaningful, although they are yet to be understood in relative terms.

Adapted for small families, households with smaller numbers of people

Undoubtedly, there has been a rather strong and sustained trend towards households with smaller numbers of people in developed countries since the 1960s and 1970s. We can quote the following examples:

- In the US State of Maryland, the average size of households dropped from 3.25 in 1980 to 2.61 in 2010 (Pew Research, April 22nd 2011). The same source states, however, that since 2010 no further decline has been observed.
- A US census showed the following numbers nation-wide: from 5.8 persons per household in 1970 down to 2.62 persons in 2005 (Jiang and O'Neill, 2007).
- A UK study showed the following numbers: from 3.1 persons per household in 1961 down to 2.4 in 2009 (Social Trends, 2010).
- A US article from June 24 2011 makes the following two further observations, which support the above trend observations:
  - First, the share of households in the USA with children dropped from 36% in 2000 to 33.5% in 2011.
  - During the same period, and only superficially related, the USA has more households with dogs now (43 million) versus households with children (38 million).

    The latter observation, which is a fairly visible reality in our cities today, has important effects and influence on product and packaging developments in the food industry, which will be discussed and analyzed in later chapters (Chapter 3 and others).
- In Japan, the average household size was mostly unchanged during the period between 1920 and 1950 at around five persons per household. In 1970 this number dropped to 3.41 and is presently (since 2010) at around 2.42 persons per household, very similar to the UK.

Households and Household Members

Year	Households (1,000)	Average annual rate of increase (%)	Household members (1,000)	Members per household	Population (1,000)	Average annual rate of increase (%)
1970	30,297	a) 3.00	103,351	3.41	104,665	1.08
1975	33,596	2.09	110,338	3.28	111,940	1.35
1980	35,824	1.29	115,451	3.22	117,060	0.90
1985	37,980	1.18	119,334	3.14	121,049	0.67
1990	40,690	1.38	121,545	2.99	123,611	0.42
1995	43,900	1.54	123,646	2.82	125,570	0.31
2000	46,782	1.28	124,725	2.67	126,926	0.21
2005	49,063	0.96	124,973	2.55	127,768	0.13
2010	51,842	1.11	125,546	2.42	128,057	0.05

a) Annual rate of increase between 1960-1970

Figure 1.2 Household sizes and household members in Japan, 1970–2010. (From Ministry of Internal Affairs and Communication, 2011.)

Figure 1.2 depicts these numbers in more detail and gives a good insight into the population evolution and the food and food-packaging requirements that come with these.

All the above suggests that smaller households with fewer members need smaller numbers of portions and the packaging that goes along with this trend towards smaller items. It is suggested that packaging unit size should be smaller, organized in such a way that a one- or two-person household is well served, meaning that the packaging size should be suitable for one or two people. Packaging should therefore be adapted to smaller portions and it would be desirable that packaging is lightweight and does not add too much weight to the product. The latter is especially important when it comes to shopping and carrying shopping bags to and from the car or up the stairs or to any other destination. Especially in a country like Japan, where the pantry is non-existent and fridges are small, the role of personal food storage is played by the ever-present neighborhood stores that one can find every few hundred meters, even in residential areas. Shoppers go there several times a day to specially buy refrigerated or frozen products and carry them home.

Safety

Food safety in general has become one of the most crucial aspects in the food industry. It has actually become more important than costs, ex-factory costs, margins, taste, marketing and sales, and everything else in the value chain. All these other aspects can be perfect, but if food safety is not guaranteed and delivered to the consumer, then nothing else matters. Apart from safety that has to be built into any food product through the hurdles of acidity, heat treatments, aseptic filling, or similar techniques, packaging plays the ultimate protective and preserving role.

A few features of safety in packaging have to be remembered. Packages have to be safe and protective even if there is careless mishandling in the supply chain. This is not an easy task, as this means that the packaging material and design have to be made in a way that takes abuse in the supply chain into account and that obviously means higher costs for the packaging solution. We therefore walk a fine line as to how much additional safety through packaging strength, barrier properties, or similar features is to be built in without adding too many costs. This is where good, no great, design, mostly structural, meets with great material knowledge and intimate knowledge of the supply chain and what types of abuse could happen.

Another important feature is not too far away from the one just discussed, but concerns the end consumer. Materials not only have to be safe on their way through the supply chain, but have to be equally safe, friendly, and good to handle, touch, and dispose of. Haptics are of utmost important, the consumer needs to have a good feeling when touching a packaged food; it should feel strong and good in his or her hand and should instill a feeling of safety and solid preservation.

Finally under the heading of safety, food packages have to be absolutely tamper proof. This is not always easy to obtain, but again is a function of packaging material quality in combination with great design. To some degree, tamper-proof packaging is also linked to eliminating or rather reducing the danger of counterfeiting, although this is a separate topic and we will discuss this later, in Chapter 3. There are several ways to render food packages tamper proof and they are very product dependent and differ slightly in their execution from flow-wrapped candy bars to bottled beverages to breakfast cereals. Common to all is a feature that can easily be detected by the end consumers and which can assure the user that the specific packaged good has not been opened before, by anyone.

We will not discuss criminal tampering by injecting poisonous substances with hypodermic needles through flexible or soft packaging into the food product. These incidents are very rare and can hardly be prevented if some criminal mind really wants to perform such an act of tampering.

Easy to open

Easy-to-open packages should be a natural given in the packaged goods industry, but, from many years of experience and many years of technical and design struggles, it clearly is not "natural" at all! Why is that so? Contrary to safety, which is a "killer" if it is not right, easy opening is not a must, not an absolutely necessary feature, maybe not even "nice to have" in the eyes of developers and marketers in the food industry. It is a totally wrong position, but it is a reality. Every marketing person will ask for easy opening and technical packaging experts, together with designers, will find really good solutions, often not even difficult to introduce into the packaging, but maybe costing one cent more per pack and that's a NO.

Consumers put a big YES on the need for easy opening and it is very interesting, even amusing, to watch some video ethnography of, for example, Chinese homemakers sitting in a room and being asked to open a series of different packaged food products of all sorts. If it wasn't a serious matter, it would almost be comical. The author has personally seen such a film and the results were really striking. In the absence of easy-opening devices, one could see a lot of packaging "mutilation" going on, with or without tools such as scissors or knives.

Let us list a few key, necessary sub-features here. First, we recognize the need for an easily detectable location for the opening. Sounds logical, sounds like a must, and yet, it is not consistently executed. In the absence of good visual detectability, even the most efficient easy-opening feature will not function as desired. Why is this not done consistently? Again it is cost that is a driving factor and, ideally, the marketing group will tell the technical and design group to make it happen, but at no extra cost. It's a mantra that we will hear again and again throughout this book, and it should be taken for what it is: a reality and not an excuse, and an opportunity for designers to work their magic and make things happen at no extra cost.

However, there is something else that the technical and design group needs to learn to apply: it's the notion of "holistic costs," i.e. what does it really cost the company to lose the sales they could make, had they applied common sense and logical solutions to packaging that may cost a bit more, but largely offsets the extra cost by increased likeability of the product and consequently increased sales, profitability and margins.

The second important point is for the consumer not to have to apply much physical strength to open a packaged food product. If this is not done right, it is one of the most important deterring factors when it comes to packaging. One could jokingly say that the lowest-calorie product is the one that I cannot access because its package resists opening, but this is definitely the wrong approach to low-calorie products. Consumers will just not buy again, which is bad news for the marketing and sales people, and blame will quickly go to the designers and technical people who have come up with "such a bad packaging solution."

Remember, "It's the costs, stupid … " No excuse, just a reminder and again a great opportunity to find ingenious solutions.

A pretty similar aspect is easy operation, meaning that once the consumer is able open the pack, he or she can easily handle the package itself, possibly heat up the product in the package (before or after opening), and, finally, the last aspect relates to easily picking out the contents of the package, pouring them out, scooping them out, handling them easily and logically in any necessary way.

Easy to see

The package of any industrial food product represents a unique opportunity to depict and describe its contents in the best possible way. Marketing has certainly realized this, and designers, most of the time, do a fine job in creating and executing good visual content on food packages. However, a few points may be worthwhile mentioning. First, any graphics such as letters, numbers and logos should be executed at a suitable size so that the entire graphic display can still be read, even by those who might have to scramble for their glasses. Second, especially with regards to the aged consumer, suitable color combinations should be applied; this is not a cost factor and designers, together with marketing, should find the best possible solutions based on experience and know-how of experts in this field. Last, easy visual discrimination is a must, the consumer *must* be able to make something meaningful out of the graphic landscape on a food package.

All these elements present unique opportunities to make "my product" really stand out and be appealing to the eye of the consumer, and therefore are the most obvious to execute in the best possible way and a fantastic playground for designers, technical packaging experts, and marketing to collaborate in a very close-knit fashion.

On a similar note but in a different industry, the Hanes Company (makers of underwear) in summer 2013 started to release a TV commercial that talked about "Go Tagless". It emphasized the advantages of eliminating any tag (label) from the interior of their underwear products, because it is a recognized nuisance and bothers the wearer. It is difficult to judge whether such a thing could be possible for food products, at least with today's legislation in place, but it might be an interesting thought and worth further reflection.

Easy understanding of label claims

This is a frequently grossly overlooked area and the wrong approach can very often lead to very unsatisfactory packaged food products. Although the graphic landscape on the package may look very decent or even appealing, explaining in simple words what the various claims and numbers on the food label mean is not an easy job. This task is most of the time left to a group of people that has no or very little knowledge of good communication and therefore often fails the task. Consumers are simply overwhelmed by the complexity of the message and, "less is more" is an absolute must, i.e. explain in as simple as possible words what the message and claims are all about.

Again, this is not an easy task and requires some communication mastership, or, like the former British Prime Minister Winston Churchill (and several others to whom this quote is attributed) once wrote: "If I had more time, I would have written a shorter letter." The art of being short and concise is a difficult one and it may be suggested that marketing people in a food company should hire the services of daily newspaper journalists, who have a great, sometimes too great, mastership of being short and concise, out of sheer need and space limitations.

Actually, this whole aspect can be summarized in one simple message: "The medium is the message," (Marshall McLuhan, Monday Conference on ABC TV, June 27 1977). The medium is the food package and it should carry the message of what it is all about in the simplest, most understandable way. Not an easy task, but a great challenge for communications and media designers.

Easy handling

The last concept of good packaging we have to look at is easy handling. The handling part covers all areas of the supply chain, starting from the end of the line to the mouth of the consumer. The following elements are of importance and some or all of them overlap in their desired and required functionalities.

Good preservation is an important requirement for good food packaging. The packaged product should and must have the indicated shelf life under the proposed storage conditions. This means that the product, from the day it is put into its package, has to be stable and safe until its consumption, even if such consumption only happens on the last official day of its shelf life.

Let us bring in an example from the world of industrial confections, chocolate and chocolate related products. In a country, a market as the food industry calls it, like the United Kingdom, empiric evidence shows that a candy bar is typically—on average—consumed some four months after it was put into its wrapper. That is to say that some candy bars are eaten earlier and some much later, but the vast majority is finding its way into the consumer's mouth after around four months. In the USA, the numbers are slightly different, but only by approx. $1/2$ month, i.e. the peak of consumption is after $4\,1/2$ months, largely due to the longer distances in transport and distribution. Despite this knowledge, products have to be safe and taste good until the end of their legal shelf life.

It is clear that distribution chains largely depend on the nature of the product. Whilst confectionary products or other ambient-temperature-stable products typically have shelf lives between 9 and 18 months (in the past even longer), refrigerated products such as yoghourts, fresh cheese, and other dairy products have much shorter shelf lives (depending on the geography, between 28 and 40 days, some even slightly longer) and therefore packaging solutions, as well as distribution efficiencies have to be adapted to this situation.

In the supply chain it is of utmost importance that packaged goods can be handled easily, multi-packs and secondary packaging containing 12, 24, 48 or any other combined number of individual products, is easy to grasp and does not exceed a certain weight, making it difficult for handlers in the supply chain to grasp and carry a secondary packaged box from a production belt or a pallet. This also means that the packaging itself, whether primary around the product or secondary in a transport carton, is solid and has good mechanical stability, not only when it comes to handling, but especially when it comes to stacking.

Finally, for the end consumer it is often of additional importance that a food package is handy, can easily be grasped and held, and, once opened, can easily be reclosed, repacked, or measured when taking out part of its contents for cooking or other consumption purposes.

THE ROLE OF PRODUCT DESIGN IN THE FOOD INDUSTRY

Thus far we have linked design in the food industry only to packaging design, in all its relevant forms, such as graphics design, industrial design, and communication and media design. This is not surprising as the food industry itself sees design pretty much only linked to these areas, and therefore negates or simply forgets that there is product design already existing, or ongoing in some major industrial food products. However, there are probably many more opportunities for applying the concept of form and functionality to the food products themselves.

Let us look into two prominent examples that represent the world leaders (or close second) in their respective classes.

The first is freeze-dried Nescafé® from the Nestlé company, the most popular and most global soluble coffee in the world. At its inception back in 1938 (Max Morgenthaler, Nestlé Orbe, Switzerland, 1938), Nescafé® was a more or less uniform, more or less stable powder that was based on liquid extraction of roasted coffee beans. Before the industry "law" was created that soluble coffee could only be based on ingredients such as coffee and water, the initial hygroscopicity of the first Nescafé® was alleviated by the addition of a carbohydrate such as maltodextrose. It is clear that a powder does not really lend itself to being designed to follow the quest for "form and function," if not to say that the form is small particles and the function is good and rapid solubility but, admittedly, it's not only a simplistic, but also a rather dull view. So how can a soluble coffee powder be designed? What type of design, still following the form—function mantra, can be performed? Form meaning shape, meaning different types of particle shapes, meaning larger particles, meaning ultimately granules of a specific shape and size, still with the functionality of being perfectly soluble.

This is where the invention of freeze-dried coffee by Nestlé's continuous freeze-drying process (Nestlé Research & Development Center/Product Development Center Marysville, Ohio, USA, 1966) came into play. This process, without going into much procedural detail, has made it possible to come up with rather defined granular particles that had very little powder (fines) in the final mix and which, due to the porosity of the particles, led to excellent dissolution properties. It can safely be said that this product design was not an active act of design, led by designers, but the engineers involved in this process apparently had a very good notion of the concept of "form and function" and applied it intuitively totally correctly, especially as one of the co-inventors of the continuous freeze-drying process was a microbiologist, well versed in nature's approach to this very concept (Tom Roth, personal communication, 1992).

Staying in the field of coffee, let us look into what one could call a "hybrid design concept," namely the well-known shape of the Nespresso® coffee capsules. We call it hybrid, as this is neither a pure packaging design solution, nor a product design, but the shape, size, and material of the capsule are exactly designed to meet the product requirements of the roast and ground coffee within. In a following chapter (Chapter 3) we will discuss this in more detail. In that chapter we will also, for comparative purposes, look at good product, graphic, and communication design in nearby industrial areas, such as household goods and small appliances.

Another well-known example for product design is the original Four-Finger KitKat®, created by Joseph Rowntree, coincidentally also in 1938. The product was licensed out to Hershey's for exclusive distribution in the USA back in 1985 and Nestlé acquired the Rowntree Mackintosh company, including KitKat® in 1987 and has the ownership of the brand and sells the product everywhere else. KitKat®, after Mars' Snickers candy bar, is the second largest selling candy bar in the world. Joseph Rowntree, an equally shrewd and visionary business man, conceived KitKat® in a very distinct and designed fashion with two main criteria in mind. First, he wanted to have a candy bar that combined the full taste of good chocolate and the lightness of biting into a layered wafer, and second, he wanted to have ideal processability, where specific ideal shapes for molding and de-molding drove the ultimate geometry and ratio of the KitKat® finger. We can truly speak of very insightful product design, forcefully applied to the product shape and manufacturing, as well as eating function. Due to the fact that KitKat® was composed of four thinly attached fingers, the consumer experience was one of breaking of finger after finger, therefore giving great portionability, as well as the starting point of a fantastic marketing campaign that has lasted decades and still lasts.

Other, more mundane examples of product design in the food industry are again to be found in the confectionary area, especially in seasonal confections, such as chocolate Easter eggs (UK, Germany, Switzerland), Easter bunnies or Santa Clauses, and the like. Recently a court case was settled between the Swiss chocolate company Lindt and the German confiseur Riegelein regarding the "golden Easter bunny," in which the German side got the upper hand by claiming a long-standing and publicly perceived notoriety with consumers with regards to their golden Easter bunny, thus not allowing Lindt to claim exclusivity on their golden bunny.

We will discuss and analyze these and some more products, especially in the area of ice cream, where we also find good examples of designed food products, as well as combinations of product and packaging design.

CONCLUSIONS

In this chapter we discussed and analyzed the following areas:

- We very briefly introduced the idea of elevating the role of designers in food companies, ideally having a designer at board level.
- We did some looking back at design in general, some historic perspectives as to the importance of design from the onset of the second industrial revolution (the automotive era) and how design was perceived and used.

- We discussed some of the consequences of the next industrial revolution, which started with the invention of plastic materials and which gave the freedom to designers not only to experiment but also to exaggerate.
- We discussed design education, some global examples, and their specific strengths.
- We introduced the concept of "form and function" as a very basic, yet extremely important concept, especially in the field of industrial/product/packaging design. Nature basically "builds" all its elements (plants and animals) following this concept!
- We analyzed and discussed fashions in design and some of the mechanisms as to how such design trends are created with the examples of home design and home improvement.
- We analyzed and discussed design in the food industry, especially along the lines of packaging design and more specifically discussed the concept of "universal and inclusive design" by listing the necessary elements for good packaging solutions.
- Finally, we discussed the limited importance of functional design in the area of food products themselves.

TOPICS FOR FURTHER DISCUSSION

- Based on your observations and experience, please list examples of good packaging concepts and solutions in the fast-moving consumer goods industry, especially the food industry.
- Discuss in more detail the content of Figure 1.1 and list elements that are missing, according to your own observations and experience.
- Discuss fashion trends and how they are created, and how a product developer, a scientist, an engineer, and a designer could possibly play the game of fashion creation equally successful.
- Can you come up with further specific examples of good design in food products? Please list such examples.
- Do you have or know of creative solutions to solve the improved functionality versus cost riddle? Please list and discuss good examples from your environment and don't get too much hung up on a long list of negative examples, if possible.
- If "all is flowing," as is stated in the introduction, and if nothing remains still, how can we make this work for design in the best possible fashion? Probably a good occasion to discuss this over a good glass of wine, or two.
- The authors would appreciate receiving feedback on any of the above questions from you. Thank you!

REFERENCES

S.G. Lewin (1991). *Formica and Design: From the Countertop to High Art*, Rizzoli International Publications, New York, p. 18.
L.L. Jiang, B.C. O'Neill (2007). Impacts of demographic trends on US household size and structure. *Population and Development Review* **33**(3): 567–591.
Ministry of Internal Affairs and Communication, Statistics Bureau, Director General for Policy Planning (Statistical Standards) and Statistical Research and Training Institute of Japan (2011). http://www.stat.go.jp/english/index.htm
Social Trends (2010) **40**, 13–16.

2 Design: from object to process

I want people to understand that design is so much more than cute chairs—that it is first and foremost everything that is around us in our life.

Paola Antonelli, MOMA, New York

THE EXPANDING ROLE OF DESIGN

It is inevitable that at any cocktail party or social event that designers go to and get asked the question: "So what do you do?," the answer of, "I am a designer," raises eyebrows, and the next question that typically follows is "Oh, you are a fashion designer?," or "I have a friend that is an artist too." The designer will then respond: "Well I'm an industrial designer—I've designed cars, toys, furniture, and accessories." The conversation then turns into one focused on styling, or reeling off a list of favorite cars, or an invitation to share their complaints, or great ideas on features they wish were in their products. Then the shift is to: "So you studied engineering, right?" And the conversation turns into a lecture about what design really is—doing research to understand users needs and desires, translating that research to come up with lots of different ideas, solving problems, discovering opportunities to apply emerging technologies, collaborating with other disciplines, make strategic business decisions, and of course visualizing all of this throughout the entire process—drawing, modeling, and prototyping—all of this—in order to achieve the goal of introducing innovative products that the world has not yet seen.

A common preconception of design is that it is simply the act of making things look pretty, and that the designers come into the process of developing a product after an idea has been generated by the marketing department, or a technology has been developed by a scientist, or after the product has been engineered. Unfortunately, that process does still happen in quite a few companies, especially companies that do not have an understanding of the complete value design brings to the table. Of course the primary function of design is to bring ideas to life, to visualize ideas on paper or a screen, to bring three-dimensional form to ideas through model-making and prototyping. Those are core and differentiating

Food Industry Design, Technology and Innovation, First Edition.
Helmut Traitler, Birgit Coleman and Karen Hofmann.
© 2015 John Wiley & Sons, Inc. Published 2015 by John Wiley & Sons, Inc.

skills that all industrial designers must have—the ability to draw and to make (and from a designer's perspective, that craft is what typically separates designers from their partners in marketing and engineering).

While that are several forms of the practice of design, the most commonly known are graphic design, advertising, fashion, interior design, and architecture. The practice of industrial design, including product design, transportation design, and interaction design more often than not gets mistaken for the field of engineering. For industrial designers to have a basic understanding of engineering can be beneficial when it comes to designing functional aspects of a product, but really the differences are vast. While engineering typically is focused purely on making things work, the designer is responsible for creating the entire user experience—from the initial concept of the product idea, to production, to the shelf of the store, and for the life of the product.

The focus of this chapter is through the lens of industrial design and the expanding role that this discipline is experiencing—from designing objects to process and systems, to services and experiences. Over the last century there are "legends" in our industry who have designed world-famous iconic products, furniture, and automobiles—from the mid-century greats like Raymond Lowey and Henry Dreyfuss, to today's contemporary rock stars like Marc Newsom and Jonathan Ive. Architects have become "starchitects" with their bold and ambitious buildings, like Frank Gehry and Zaha Hadid, and have also extended their design language and vision into home products, jewelry, and clothing. These designers and many others over the last several decades have shaped our culture and also brought an awareness of the profession of design to the world at large.

Specifically over the last decade the processes that make up the practice of design have been introduced into the world of business, strategy, and service sectors as an innovative process called "design thinking." Legendary design firms like IDEO, along with their leaders, David Kelley and Tim Brown, have demonstrated the power of design going beyond bringing beautiful artifacts into the world. Authors and educators like Roger Martin, Dan Pink, Malcolm Gladwell, and Thomas Lockwood have introduced techniques and case studies that describe how design thinking is transforming the way companies innovate, how corporate culture is shifting, and even playing a role in development of governments. The process of design is one that is being embraced by corporate culture, not just to design products, but to also play a central role in leading innovation within the organization. This chapter will highlight several examples of how companies leveraging design and design thinking are transforming their culture, how they do business, and what they put out into the world.

WHY NOW? DRIVERS OF CHANGE = THE INDUSTRY SHIFTS + DESIGN EXPANDS

Along with the emergence of design thinking in business culture, there are many reasons why the design profession and the roles of designers have been significantly expanding over the last several years. Never has there been a more dynamic and opportune time for the industry to leverage design in its thinking and in its processes to innovate.

New platforms / new options

No doubt the internet and social media have played a significant role in this expansion of design. Myspace, Facebook and Twitter opened the world up to sharing ideas—in real time—and for members of those communities to get instant feedback (or "likes"). Combine those social enablers with online retail and online gaming and now you have a new economy based on crowdsourcing ideas and the ability to use consumers (otherwise known as members or backers) in the development of products or services that cater to personal interests, wants, needs, and desires. New online platforms like Kickstarter, Indiegogo, Etsy, and Craigslist have empowered designers and creators to take their ideas direct to the marketplace, bypassing the middleman. Ideas that once sat in designers sketch books and never saw the day of light because of the companies, manufacturers, and producers that the creators relied on to invest and bring those ideas to market are now being brought to life through crowdsourcing platforms. We are deep in the "Kickstarter culture" as we are seeing record-breaking success among designers bringing their products to the market.

A couple of headlines from the phenomenon of Kickstarter:

Scott Wilson/Minima–LunaTik

In November 2010 designer Scott Wilson launched watch kits for Apple's iPod Nano on Kickstarter, well before the crowd-funding platform was being utilized by designers. "Tik-Tok" and "LunaTik" wristbands transformed the iPod Nano into a wristwatch. The fundraising goal was set at $15,000 in 30 days to get the products into production. Little did Scott or the design community know how disruptive this project would become. In three days, the fund received more than ten times the original goal, reaching $180,000. In less than a week over 5,000 backers contributed nearly $400,000 to TikTok and LunaTik, and at the end of the standard 30-day fundraising period, the project reached a Kickstarter record-breaking $1 million. Instantly, LUNATIK became a global mobile and digital life-products brand built on this crowdsourcing model. The brand that started of as a bit of an experiment for the pioneering Scott Wilson ended up paving the way for the design entrepreneurs and product developers around the world to take their ideas right to the consumers.

Julie Uhrman, Yves Behar, and the Ouya game console

Founder and CEO Julie Uhrman teamed up with industrial designer Yves Behar to launch Ouya—an affordable open-source gaming console running on the Android operating system. Ouya's Kickstarter fundraising goal was raised within eight hours and holds the record for best first day performance of any project hosted to date. The project became one of the most quickly funded projects on Kickstarter to raise more than a million dollars. In June of 2013 the Ouya was released to the public for $99 and is now sold in major retailers.

Kickstarter and other platforms have not only generated options for direct to consumer channels, but have also opened up career options for the designers, despite the current challenging economic landscape. The typical career trajectory for designers is evolving and diversifying, perhaps because of the entrepreneurial spirit that economic downturns spawn.

The traditional path of going to design school, getting the degree, and going to work for a large corporation or a consultancy, or perhaps freelancing in the hope of launching a design business, has dramatically changed. Students graduating from design school now may already have a product on the market thanks to a successful Kickstarter campaign. In fact, some students may not stay in school to earn that degree because their new venture has been so successful. Of course, the majority of design school graduates are going to work for a company during the day, but after hours, at night or on the weekend, are developing their own ideas to launch or license. Regardless, these new platforms and options are contributing to a compelling entrepreneurial economy that many are benefiting from—from inventors with a shop in their garage dreaming of the next big thing to venture capitalists seeking innovative ideas to invest in. The role of design is playing a vital part in bringing revolutionary ideas to market and developing an entrepreneurial mindset to move into the future with.

Speed to market / direct to market / new retail models

Along with the crowdsourcing platforms, the speed at which products can be manufactured (at reasonable costs) as well as technologies like 3D printing and other robotics, are changing the face of how products of all scales are getting made. Over the last decade the rise of rapid manufacturing in China and the entire southeast Asia region has enabled both global brands and boutique "micro-brands" to introduce their products to the world at a record-setting pace, outsourcing production directly to low-cost Asian suppliers. We will not get into the issues around quality of goods and politics of labor laws in this chapter, but the growing "backlash" from outsourced manufacturing is getting multinational corporations to balance their global manufacturing portfolio with local hubs, just-in-time factories, and cottage industries. Innovative manufacturing models are just one part of the equation in the speed to market, or direct to market equation.

For example, visionary designers and partnerships located in Hong Kong, Shenzhen, and small design offices around the globe are proving that mass production is not the only way to introduce new products to the market and grow a brand. Roger Ball, author of *Design Direct* describes in his book how these "micro-brands" are bypassing corporations and traditional distribution channels to reach customers directly over the Internet.

Game-changing entrepreneurs like Elon Musk are rewriting the rules of how products as complex as the Tesla electric vehicle are being designed, produced, and sold—with a vision of the entire experience. Tesla is a technology- and design-driven company. In 2008, Elon brought chief designer Franz von Holzhausen (former chief designer at Mazda, GM, and Volkswagen) on board to lead a design team to visually define the Tesla experience—giving form and design language to the vehicles, Tesla stores, and charging stations. At the moment, Tesla is the brand that has completely redefined what a car and ownership can be—a luxurious alternative electric vehicle, manufactured locally, sold direct to consumers online and via Tesla experience boutiques, lifetime servicing for free, along with battery swapping, and developing a network of Supercharger stations for long-distance trips anywhere in the nation. Elon's strategy has caused havoc with the National Automotive Dealers Association, bucking the traditional way cars are sold. Instead of having buyers deal with haggling salesmen, Tesla has curated a new buying experience that feels more like a boutique furniture store. The "sample" cars are in the middle of the store with walls displaying options in interior components, color and materials choices, and information on the vehicle technologies. Prices are not negotiable and the clerk is there to help buyers make their choices in customizing their vehicle, which will be ordered online, manufactured in northern California, and delivered to their residence when complete. Every detail of the Tesla experience is defined by design, not just technology, making this brand one of the most disruptive innovation benchmarks in automotive history.

Open innovation / systems innovation

On the corporate level, there has been a significant shift from keeping innovation protected and in-house to one that reflects a participatory culture. Multi-national behemoth companies with multiple brands are recognizing the power of the creative collective, otherwise known as co-creation, and understanding that innovation does not come from just one way of thinking or abiding by set improvement processes within organizations. Proctor & Gamble's Connect & Develop program boasts on their homepage: "Could your innovation be the next game-changer?" P&G's open innovation program began in 2001 and has seen numerous innovative ideas born from this participatory platform. The Tide Pods, Crest White Strips, and the Bounce Dryer Bar are great examples of successful products born from the partnerships established by P&G's Connect & Develop program. This program will be discussed in more detail in Chapter 11.

Other big organizations are seizing on the participatory design and innovation trend and joining forces to take on some of the worlds most urgent challenges and support work in creating sustainable futures. The LAUNCH global initiative began in 2010 with the motto: "Collective Genius for a Better World." Founding partners including Nike, NASA, USAID, and the Department of State have hosted a series of forums showcasing and supporting visionaries with ideas that have potential to have a major positive impact on society. Through various open-source challenges, LAUNCH identifies and accelerates innovations that could change the world. Topics have included water, health, energy, waste, and systems, defined by a team of thought leaders and representatives from each of the founding organizations. This open-source innovation model demonstrates the potential for governments, global development organizations, corporations, universities, inventors, and investors to collaborate in powerful ways that the world has not witnessed before.

Related to the open innovation and systems innovation models is competitive innovation modeled by organizations such as the XPRIZE, a non-profit organization that creates competitions around their motto: "Making the impossible possible." XPRIZE is a global incentivized prize competition model that rewards innovators—individuals, companies and organizations—with radical breakthroughs in areas such as space exploration, energy systems, mobility, and health, and well-being. In 2004, founder Peter Diamandis offered a purse of $10 million to the winners of the Ansari XPRIZE for Suborbital Spaceflight to Mojave Aerospace Ventures who developed the Space Ship One, a three-passenger vehicle that can fly 100 kilometers into space. This competition had 26 teams from around the world investing more than $100 million in pursuit of the $10 million prize. It could be said that this prize was the catalyst for entrepreneurs like Sir Richard Branson to invest in the emerging personal spaceflight industry. Since the Ansari XPRIZE, a number of challenges, past and present, are demonstrating the power of this competitive innovation model. Great ideas that advance the state of humanity can come from anyone, anywhere and can be rewarded.

Creative economy / sharing economy

In 2006 author Daniel Pink released his book *A Whole New Mind: Why Right Brainers Will Rule the Future*. His premise, made several years ago, launched the notion of the creative economy and can be seen manifested in the drivers of change stated in this chapter. According to Pink, creativity becomes the competitive difference that can differentiate a commodity-driven world of consumer abundance, outsourcing of manufacturing, and automation of services. Pink states that we are moving from the information age of knowledge workers to the conceptual age of creators and empathizers. The book emphasizes skills typically belonging to the right-brained folks of the world that will rule the future of the global economy: moving beyond function and engaging the senses, including storytelling in products and services, engaging emotion and intuition, big-picture thinking, and even bringing humor into business and products.

While the creative economy continues to expand, especially in times of economic crisis, another model of business is booming, especially with the generation of consumers who have embraced technology-based peer communities, and approach consumerism and capitalism with a sharing versus owning mindset. *What's Mine is Yours: The Rise of Collaborative Consumption*, a book by Rachel Botsman and Roo Rogers, notes the enormous success of marketplace companies like E-bay, Craigslist, car sharing companies like Zipcar and RelayRides, peer-to-peer accommodation like Airbnb, peer-to-peer task assignments like TaskRabbit, and new ventures like Uber and Lyft, who are redefining car service. Botsman and Rogers suggest models of consumption that emphasize usefulness over ownership, community over selfishness, and sustainability over novelty. For those that are citizens of these internet networks, individual ownership is trumped by affordable products and services, easy maintenance, and a feeling of being part of a social community that shares those ideals.

Maker culture / hacker culture / DIY / new craft

The next two drivers are inter-related, but deserve specific attention as one is about a mindset or culture movement and the other is about the enabling technology. The "maker culture" that is thriving today stems from the do-it-yourself or DIY movement that began around a decade ago. MAKE magazine and other publications, both online and in print, became a platform for crafters and "hackers" to share their projects and inventions, from repurposing objects into furniture to wiring electronics and making robots. The community grew as the publication launched "Maker Faire," gatherings around the country of like-minded people who led DIY lifestyles and want to celebrate their arts, crafts, engineering, and science projects . Since the first Maker Faire in San Mateo in 2006, these faires have now take place in cities all over the world. Scientific and electronics exploration that used to be found only in places like MIT Media Lab have permeated the general public because of Maker Faire and inspires people of all ages to play, imagine, and invent. With the availability of Arduino, single-board microcontrollers and open-source software used to make electronics that are both accessible and affordable, the DIY movement continues to flourish.

Around the same time that MAKE launched, so did Etsy, an online marketplace focused on handmade or vintage items, arts and crafts supplies, and a blog for sharing DIY techniques. Instead of going to local craft fairs to find handmade gifts, Etsy made it possible for makers to have personal storefronts that could be accessed by buyers anywhere in the world. The craft market has exploded as these makers have leveraged this e-commerce platform and society at large has developed an appreciation for the handmade in a world that is so technology based. Currently, there are over 30 million users registered on the site with projections of 1 million sellers registered on Etsy with over $1 billion in total annual transactions by the end of 2013.

While there seems to be a rebirth of arts and crafts in the United States and around the world, technology is permeating the once made-by-hand-only processes. Similarly, there is an emergence of artisan approach and tools influencing mass production. A recent project at Art Center College of Design and TAMA Art University in Tokyo explored the topic of "Future Craft." Students spent a semester researching and exploring traditional hand-crafted methods of textile weaving, glass blowing, ceramics, papermaking, woodworking, metallurgy, and the like, while at the same time mixing these traditional crafts with new tools, 3D printing, and other technologies, an example being a design student's development of a lounge chair inspired by the symbolic layering and draping of the traditional Japanese kimono. The "KASANE" lounge chair form was derived from the unibody shape of the kimono transforming into an organic shape with a few simple folds and cuts. 3D modeling software was used by the designer to create a series of wireframe chairs to test the ergonomics and molding process that a dozen layers of different colors of felt had to respond to, the final product reflecting the spirit and aesthetic of an ancient tradition enabled by a technology to make it reality.

On a mass production scale, Nike's "Flyknit" performance running shoe is a fine example of industry rethinking traditional cut-and-sew production methods by combining old technology with new technology. The Flyknit is a shoe fabricated without seams and crafted in one layer by a machine knitting its structure based on input from a software program that has translated the original design into data. The result is a lightweight, breathable, high-performance running shoe used by professional athletes, everyday runners, and those the enjoy wearing shoes with a unique aesthetic. A by-product of this revolutionary process of making a shoe is the sustainability factor. While the goal was to give athletes the best fitting, most breathable, and most lightweight shoe possible, because of these attributes, the Flyknit is also an example of a product that uses the exact amount of materials needed to perform, leaving behind less waste, less adhesive, and a lighter weight product to ship. The future of applying new technologies to traditional craft and making processes is in its infancy, as innovative companies like Nike experiment and succeed with products like the Flyknit.

3D Printing

The final driver of change explained in this section is 3D printing. It will remain somewhat brief, as there is a proliferation of companies, entrepreneurs, and inventors taking advantage of this trend manifesting and revolutionizing how ideas are being made. There is no other technology that has democratized design like 3D printing, making accessible to anyone with an idea, an internet connection, simple software, and funding to produce just about anything they wish to. Companies like Dutch-founded and New York-based Shapeways enable users to upload design files and receive back 3D printed objects in the mail just days later. As of a year ago, Shapeways had uploaded, printed, and delivered more than one million user-designed objects. The creative economy is thriving, as fashion designers are printing futuristic clothing, as bakers and chefs are printing unique shapes of foods and delectable pastries, and product designers are producing lamps and accessories with intricate geometry that could only be produced through 3D printing.

While most companies represent the incredible positive benefits of the creative economy and 3D printing, there is also a "dark side" to be conscious of and, at some point in the not too distant future, there is sure to be regulations and laws around "ethical making." 3D fabricators have recently acquired data on OEM parts and are printing replacement parts or recreating entirely new "pirated" products. An example is the DeWalt drill, where a maker printed an entirely new casing. Situations like this are sure to lead to intellectual property infringement cases. And of course the most controversial 3D printing issue is the online sources for printing guns and other weapons that typically would require permits. Online digital publishing companies like Defense Distributed provide the data for anyone to print their very own weapon.

Beyond the creative economy, 3D printing is impacting social innovation as well. Bespoke Innovations combines industrial design with orthopedics and a mission to bring human solutions to people with congenital or traumatic limb loss. Most prosthetics in the recent past, while well-engineered and technically functional, are aesthetically crude, machine-like, and not personal whatsoever. Bespoke Innovations marries function with fashion, delivering custom-made prosthetics that are designed to perform and enable the user to emotionally connect with their new limbs. The prosthetic shifts from being one-size-fits-all, ordinary medical device to a form of personal expression shaped by individual need and taste. We have only just begun to imagine how technologies like 3D printing are going to transform our daily lives.

BEING DESIGN DRIVEN: ICONS AT THE INTERSECTION OF BUSINESS AND DESIGN

These drivers of change are opening doors for new economies to emerge by enabling just about anyone in the world to creatively contribute to these participatory business models. While the above tools and platforms are revolutionizing the business, access, scale, and impact of design, the process of design remains similar. Large companies like Apple, Nike, and BMW have demonstrated what it means to be "design driven" through the execution of their product and brand experiences. And companies like Proctor & Gamble and IDEO have leveraged "design thinking" to transform how business is being done, along with strengthening their brand experience. What does being "design driven" really mean? There are four threads that run through organizations that are truly design driven:

Human-centeredness—having deep empathy for the user and articulating their unmet needs in product and service interactions.
Holistic thinking—considering the entire product and brand experience beyond the object itself, from manufacturing to retail and consumption to service and end-of-life.
Problem solving—aesthetics to follow need, not arbitrary styling or decoration without purpose.
Partnership - commitment to design and business sharing a mindset and making decisions together.

The companies listed above, along with other design-driven organizations, authors of business innovation books, and the media, have created a buzz around design thinking over the last decade. It seems like it is a relatively new process for innovation, but design-driven business is not new—there are icons in the design industry from the middle of the last century that have led the way for designers to engage with business and for business, and society at large, to benefit from design. One of these pioneers is Dieter Rams.

"Weniger, aber besser" (less but better), otherwise known as "less is more," is the mantra for the legendary industrial designer Dieter Rams, former chief designer at Braun. The heart of his message is really to think about the impact products have on our society and the responsibility designers have to design for a more human environment. Rams grew up in Wiesbaden Germany in the 1930s and 1940s. Influenced by his grandfather, who was a master carpenter, he studied architecture and interior decoration at the Werk-Kunst-Schule in Wiesbaden. In 1955, Rams joined Braun, a German corporation that pioneered state-of-the-art radios and audio equipment. In 1961 he became the chief of design, spending four decades developing an array of iconic products, from radios and record players to coffeemakers and shavers. Through a close partnership with the CEO of Braun, Rams was able to seamlessly integrate his approach to design into the ethos of the Braun brand. "The 10 Principles of Good Design", first presented by Rams in 1976, serve as a foundation for human-centered design today and has had a profound influence on the design industry and design education. The Principles are not based in theory, but are the result of decades of work.

His approach certainly permeates our design culture today—the most obvious being Apple, and evident in the iPhones, iPads and iPods that so many of us carry. Jonathan Ives, head of design at Apple, attributes his own design work to Rams' austere but user-friendly aesthetic. Ives was inspired by Rams' approach to creating surfaces that were bold, pure, perfectly proportioned, coherent, and seemingly effortless. Rams believes that the world is filled with "visual pollution" (emphasized by Rams in his ninth Principle), and we are in need of more harmonious product experiences as society grows more complex. The key to good design starts with a human-centered approach. One of Rams' many philosophical statements: "You cannot understand good design if you do not understand people; design is made for people. It must be ergonomically correct, meaning it must harmonize with a human being's strengths, dimensions, senses and understanding."

According to Dieter Rams, good design also manifests with good relationships with business—and is essential for companies to innovate, and have success and relevance in the marketplace. Braun was a "design-driven" company long before that term and prac-tice gained traction in today's industry. Rams frequently credits his success to the fact that he worked directly with the CEOs at Braun, the brothers Erwin and Artur Braun, early in his career—a revealing parallel to designer Jonathan Ive's relationship with visionary CEO Steve Jobs. Today, the icon for this philosophy of good design is Apple and the rela-tionship that Jobs and Ives shared manifested in a harmonious vision of an entire brand experience—from the product and the packaging, to the retail and service design, to the digital and physical experience, every touch-point being considered in the ecosystem of the user experience. As our world becomes more and more complex and the desire and need for purity and simplicity is greater, so is the relevance of Dieter Rams' "less but better" philosophy. Rams states that designers have unique roles and valuable creative skills along with senses of curiosity, empathy, determination, and optimism, that ultimately can have positive impact on companies and society at large. Because of these skills, designers and design-driven companies have a great responsibility to shape a sustainable future that keeps this ecosystem in mind.

THE VALUE OF THE DESIGNER: A NEW MINDSET

As great industrial designers were being recognized by industry and society in the mid-twentieth century, labels ensued. The advertising industry cast an image of handsome sophisticated men in tailored suits working in modern offices, smoking cigarettes while creating campaigns and sipping martinis with clients over liquid lunches. In the AMC television series "Mad Men," this image is typified by the lead character Don Draper. In the late 1980s and 1990s, car designers, product designers, and architects became rock stars in a similar way that fashion designers were introduced to the world—seductive, sultry, well-coifed, and typically wearing black. While their cool image was embraced by themselves, and the press, it was challenging for them to fit inside a traditional corporate culture. They were the crazy creative ones that could be found in the studio down the hall, kept away from the sightlines of executives and customers.

It was not until the designer's hero, Steve Jobs, delivered to the world a new manifesto that elevated the value of design that corporate culture began to appreciate the value of design, and began tearing down the walls (and cubicles) between design and engineering, marketing, sales, etc. Even the C-suite,

Apple's "Think Different" campaign created in 1997 with "The Crazy Ones":

Here's to the crazy ones. The misfits. The rebels. The troublemakers. The round pegs in the square holes. The ones who see things differently. They're not fond of rules. And they have no respect for the status quo. You can quote them, disagree with them, glorify or vilify them. About the only thing you can't do is ignore them. Because they change things. They push the human race forward. And while some may see them as the crazy ones, we see genius. Because the people who are crazy enough to think they can change the world, are the ones who do.

Apple Inc.

And yes, the walls are coming down as leaders of organizations are realizing that innovation does not happen in isolation. While there are still companies that develop products by having designers toss their ideas over the wall to the engineers to go make, or the marketing folks tossing their research reports over the wall to the designers to come up with the next best widget, it is more common now to see the cubicle give way to the open-air collaborative studio. Creative environments breed innovation. The world is moving too quickly to work in silos.

So what is the catalyst for collaboration between the disciplines? How can companies bring their executive, marketing, engineering, and creative teams together to deliver innovation? It's the process of design—or design thinking—its not just about designers' creative and technical skills … its about a process that enables multiple perspectives and stakeholders to come together and create a unified vision. Designers are collaborating with other disciplines like never before and leveraging their process to facilitate the dialogue between the disciplines. The design process encourages the convening of multiple perspectives, and designers are taking on new roles that look very different from the iconic cool-imaged designer of past. Instead of emulating Don Draper, today being a designer looks more like the conductor and violinist Gustavo Dudamel orchestrating a symphony. Dudamel is the music director of the Orquesta Sinfónica Simón Bolívar and the Los Angeles Philharmonic, an incredibly dynamic and passionate act to witness at work. Designers, like conductors, orchestrate a symphony of different voices, facilitate conversations amongst various disciplines and stakeholders, and collaborate with partners to enable a vision to come to life. And of course the core differentiating value of design is the ability to translate those various inputs and visualize futures that don't yet exist. Design has become a process for convening, facilitating, and envisioning complex systems.

In terms of the design-thinking process, much has been said about the "right brain" versus "left brain" approach, typified by Daniel Pink's book, *A Whole New Mind*, the right brain being creative, non-linear, expansive, and qualitative thinking and the left-brain being analytical, logical, practical, and quantitative thinking. True innovation comes from combining both sides of the brain. Without going too deep into the design-thinking process, it should be noted that the mindset and attributes the right brain designers typically possess are critical for discovering breakthrough opportunities. These include:

- *Being optimistic*—open to possibilities and imagining positive futures
- *Determined to solve problems*—regardless of how complex
- *Explorative and experimental*—leveraging an iterative process, fast-to-fail prototyping, having multiple ideas before settling on one idea
- *Having empathy*—a deep desire to understand unmet needs and desires.

The mindset of the designer, while quite different from other the mindsets found in business and engineering, is being recognized as a critical driver for innovation. In his book, *The Game-Changer: How You Can Drive Revenue and Profit Growth with Innovation*, A.G. Laffley, CEO of Proctor & Gamble, states that: "Business schools tend to focus on inductive thinking (based directly on observable facts) and deductive thinking (logic and analysis, typically based on past evidence). Design schools emphasize abductive thinking—imagining what could be possible. This new thinking approach helps us challenge assumed constraints and add to ideas, versus discouraging them."

THE ERA OF THE DESIGN ENTREPRENEUR

As described earlier in this chapter, there are specific drivers of change that are enabling a creative entrepreneurial start-up mindset to take form, and shape our industries and economies like never before. We are entering the era of the "design entrepreneur." The combination of right-brain skill sets coupled with these trends are shifting industry, economic, and social landscapes. This entrepreneurial mindset is empowering the next generation of designers to have more options and diversity in their careers. Even as recent as a decade ago, the typical career path choices for a designer coming out of school was pretty narrow—either go work for a consultancy or a corporate design team. Rarely, students would depart school to launch their own business. Those opportunities are still prevalent, but now with platforms like Kickstarter, options to license ideas, along with the abundance of accelerators and incubators, has given designers options to either commit to an entrepreneurial path or work in duality—take the day job at a corporation or consultancy and at night develop products for licensing opportunities, or prepare your Kickstarter campaign, or be part of a start-up with other like-minded designers and entrepreneurs.

Within this generation of young (and some extremely wealthy) entrepreneurs starting up wildly successful tech based companies like Facebook, Twitter and Pinterest, there is a movement of product designers riding the wave of startup culture. Art Center has a growing legacy of alumni that are examples of this movement. In 2007, product design student Spencer Nikosey had an ambitious vision of creating his own company producing high-performance luggage, bags, and accessories. A year later, as he was graduating, instead of seeking a job, Nikosey devoted all of his energy to building his brand, Killspencer, staging his own pop-up product launch event at Art Center, inviting buyers, investors, and the general public to purchase his goods. This certainly raised eyebrows, and doubts, that this ambitious young man would succeed. Five years later, Nikosey has turned Killspencer into a premium, locally made cottage brand, with a line of products produced in his workshop in Downtown LA and sold in his recently opened Silverlake boutique.

One of Nikosey's mentors is Joe Tan, one of the founders of InCase, who spent a few years after Art Center working for consultancies like IDEO before seeing the opportunity to create his own venture, capitalizing on the growing need for protection of mobile electronics. Along with other design partners, Tan created InCase, a company focused on creating mobile storage solutions and accessories for personal consumer electronics. Today Tan has moved away from day-to-day operations at InCase to launch a new company, Moreless, developing several lines of products, from personal goods to home goods, with a focus on a holistic approach to product experiences. Similar to Tan's career path from design consultancy to design entrepreneur is the highly decorated Yves Béhar, founder of Fuseproject, with clients like Herman Miller, Puma, Mini, See Better to Learn Better, General Electric, Swarovski, and Prada. The product and brand development firm emphasizes the integration of commercial products with sustainability and social good. In 2010, Fuseproject was the top winner of the Industrial Designers Society of America IDSA IDEA/Fast Company awards with 14 winning products and, in 2011, the Conde Nast Innovation and Design Awards recognized him as Designer of the Year.

What sets Yves apart from being just a designer is his commitment to social innovation and breakthrough technology products and this is evident by his position and leadership in initiatives such as the One Laptop per Child (OLPC) XO laptop, where he serves as the Chief Design Officer and as Chief Creative Officer of the wearable technology company, Jawbone. He designs the Jawbone brand, packaging, communications, and products. Behar designed the JAMBOX and BIG JAMBOX, a family of completely wireless, bluetooth compact audio speakers, and also the UP, Jawbone's app-powered health and wellness wristband. Behar's approach of positioning himself as a partner, not just a designer, is an example of design moving away from a service-based mindset to a venture-based partnership model.

What does it all mean for the business of design in corporations and the industry at large? The sibling of the design entrepreneur is the design "intrapreneur"—those creative-change agents within organizations that have the entrepreneurial spirit and aptitude to launch new ideas, new venture ideas, technologies, or business units to give companies an innovative edge in some capacity. Typically these intrapreneurs are found in small, advanced-design teams, R&D groups or dedicated "skunk works." They are the "crazy ones" who come up with fantastical ideas, in what is perceived as a sand box or playground, at most times separated from other business units or commercial production. This type of environment is critical to have—it's unfortunate that more of these sand boxes are not found in more numbers across the organization. The "design thinkers" are embedded into unlikely areas of a company, like sales, customer service, supply chain management, and even HR.

One of the most successful examples of this sandbox approach is the Clay Street project that P&G invested in a decade ago under the helm of the queen of design thinking, Claudia Kotchka, and her design leadership team. They experimented by investing in a space away from the corporate headquarters, in downtown Cincinnati, remodeling a loft in a gritty neighborhood, creating an experience that was the complete opposite of the corporate studio. The idea was to take ten employees out of their day-to-day jobs and place them in this new creative environment for three months to explore new ways of solving problems with, at that time, radical innovation methods. This huge investment with huge risk paid off, as Clay Street has hosted over 18 sessions with over 200 employees experiencing new ways of thinking, exploring, and innovating. The P&G teams that have gone through the Clay Street experience have gone on to redefine brands, create new product categories, deliver unique product innovations, and, most importantly, have inspired a creative culture inside a giant corporation, one team at a time.

DESIGN IMPACT: MAKING / MEANING / TRANSFORMING

It is important to understand that the entrepreneurial mindset that is taking shape in our culture is not limited to those that wear the label of designer, maker, artist, or entrepreneur, for that matter. In terms of histories and legacies, design has had an incredible century of innovation, with roots in the Industrial Revolution to the current blossoming Creative Economy. Today, design has reach beyond the act of bringing ideas to life or making objects. Design has processes, techniques, tools, and skills that create impact. Coming from the perspective of disciplines outside of design, how can design have an impact on your business? How do you engage with and work with designers? Where do you begin? Whether there is an existing internal creative team or a pool of design consultants, how do you get the best return on your investment? What kind of challenges are best to involve a design team? In other words, what can design do for you? While this design impact topic will be extensively addressed, with examples in Chapter 4, "Design and technology in academia: a new approach," it is essential that there is an understanding of the capacity and scale at which design is working. There are three levels at which design has expertise—exploring ideas and making things, creating meaningful ideas and relevant inventions, and transforming behaviors and how organizations do business.

Design as a process of exploration (making)

The most common way design is utilized, and most known for, is the making of things. The core of design practice is the process of exploration and bringing ideas to life represented by imagery and artifacts—things you can touch, feel, and hold. Design is an iterative process fueled by curiosity, asking questions, identifying problems and opportunities, coming up with lots of ideas and possibilities, critiquing and testing concepts, and refining those concepts into innovative and novel ideas. In its most traditional sense, design is a craft that gives form to ideas, visualized through iterations of sketching, making models, and prototyping—the process manifesting into an artifact that represents an idea or a solution to a problem. In the end something tangible, whether an object or a story or a graphic is produced. Depending on the scale and scope of a brief or a challenge—the process for exploration and making can be hours, days, weeks, months, or even years. Regardless of the project goals, what differentiates designers from their partners in engineering and marketing is their ability to execute this iterative, visual and making process, combining aesthetics with functionality and commercial viability. Designers think by making, solve problems by making, and ultimately innovate by making.

Design as a process of creating relevance (meaning)

Design as "making" and as a process of exploration has defined the profession for the last century. That process is expected and is the foundation of design, with historical examples coming from the previously mentioned design legends. In this last decade the practice of design has significantly expanded beyond simply "making" into creating "meaning" or relevance by developing products that go beyond function and aesthetics. Good design can improve people's lives through a seamless blend of functionality, attractiveness, and relevance. At the core of this is the human-centered approach, another competency that design has owned and elevated over the last several years. Design is no longer just simply and arbitrarily styling products, but defining the entire experience and creating an emotional connection between people and their products. This idea that good design can evoke emotion and elevate the perceived value of an object is best represented in design critic and author Don Norman's book, *Emotional Design: Why We Love (or Hate) Everyday Things*. Norman emphasizes that these emotional connections are made when design addresses his described three dimensions: visceral, behavioral, and reflective states. It's no longer just about form and function. Companies like Apple, Nike, and BMW have built their brands and products on delivering emotional experiences.

An emerging area of design that plays a significant role in creating relevant and meaningful products is color and material design. The relationship between humans and their products begins with "visual first impressions" that are often dictated by the surface color and materials. Depending on the emotional response, that relationship will continue through further interaction with the product, or it will be simply disregarded. We all assign emotional resonance to colors and materials, both culturally and personally. The key to research in this field is to discover how the emotional life of a product's users can be translated into corresponding colors and materials. Or, in other terms, to study how human emotion can be expressed in color and material, as well as in form.

Several years ago at Art Center College of Design, Samsung sponsored a project that challenged students to explore the expression of human emotion in colors and materials in future mobile communication experiences—specifically the cell phone. This project brief asked students to look at culture, lifestyle, enabling technologies and materials, and color palettes in order to respond to these questions: How will the "skin" of future Samsung products engage the total product experience? How would the look and feel of such artifacts trigger an emotional response? What makes a product irresistible to pick up?

This studio focused on redefining product experiences, both functionally and emotionally, by visualizing future user scenarios around global social and cultural trends that are shaping consumer behavior, needs, and desires. Based on ethnographic and material studies, students researched emerging materials and technologies to develop innovative ideas that address the skin of the product, sustainability, and dynamic color palettes, in order to develop unique emotional bonds between users and portable communication devices. The project revealed that developing color and materials should not be an afterthought in the design process. It is rather an opportunity to redefine a product experience based on developing a "skin," where color and material become the differentiator in creating an emotional connection between the product and its user. Results proved that the methods used in this project produce design language that is directly connected to the emotional values of the end user.

Design as a catalyst for change (transforming)

If "making" is where design has been, and creating "meaning" is where design is going, then what is the future for design and its impact? By no means does design lose any of the necessary making and thinking skills, as described above, but it does mean that design has greater opportunities and added responsibilities to apply its process to transform human behavior, to transform how organizations operate (Proctor & Gamble being the best corporate example), and to perhaps transform the world by taking on wicked problems around the globe. The design-making and design-thinking methodologies not only focus on a human-centered approach, but also incorporate systems thinking, strategic thinking, and perhaps in the future, influence policy-making. Design is becoming a catalyst for shifting culture. Design can tell stories, visual stories that translate complex information into simple, digestible, and memorable insights that can drive action.

An excellent example can be found in the work of Jonathan Jarvis, a media designer that has tackled the world's most wicked problems, including the financial crisis and the water crisis, by creating visual narratives that clearly communicate the mechanics and magnitude of the problems. This method has far greater impact and opportunity to ignite change then reading a report of facts and figures. Jonathan's website features a series of videos that demonstrate the power of design's ability to look at the big picture. It's hard not to ask the question—what if designers had been integrated into the financial sectors of our governments before the world reached the tipping point of its massive economic crash? Could the visualization of critical information paint a different picture for the stakeholders? The same question can be asked of the healthcare crisis currently taking place in the United States. If design had a seat at the table of our government and policymakers—could this crisis have been avoided?

Design as a process of transformation is the foundation for another emerging area of design—social innovation, social impact design or responsible design—utilizing the design process to improve human well-being and cause real change in the world through good design. Design is entering the public sector in numerous ways, with commitment—in academia, the non-profit world, and social responsibility arms of corporations. One of design's powerhouse consultancies, IDEO, launched IDEO.org, dedicated to working with non-profits, social enterprises, and foundations. Their slogan, "We're Out to Design a Better World," represents that commitment to social innovation with a focus on making a bigger impact on global poverty. Unlike the for-profit IDEO, this non-profit organization is run with an open-source model—meaning that everyone can learn with and from IDEO.org projects. The IDEO.org Fellowship Program is fostering this new generation of designers to be leaders with expertise in design thinking and social innovation, with hopes that they will then take that expertise to whatever organizations they start or work for.

Along with organizations like IDEO.org, academia's commitment to social innovation is thriving as well. New graduate programs in social design have been launching over the last couple of years. Art Center's Graduate Media Design program in collaboration with its *Designmatters* program (where art and design education meets social change) runs a Field track where students work in a real-world context, where social issues, media infrastructure, and communication technology intersect. The program takes on the ethics, politics, and practices of design in the realm of social change, preparing designers to take an active role in the creation of new models for international development and civic engagement through work in communities, institutions, governments, and entrepreneurial endeavors. Other colleges and universities around the nation have developed social design programs, as well including the Stanford d-School, The Maryland Institute College of Art (MICA) Social Design program, and the School of Visual Arts Design for Social Change program. Further examples of social design and design impact will be found in Chapter 4, "Design and technology in academia: a new approach".

THE FUTURE OF MEANINGFUL PRODUCT EXPERIENCES: DESIGN DELIVERS

As this chapter concludes, it is important to summarize the value of design and how design delivers to organizations and the world at large. As we move into the next era of innovation, a new DNA is emerging for designers. Designers are not just designers anymore. With strong partnerships, designers can be catalysts for transformation and can define a new future for meaningful product experiences.

Design creates meaningful value propositions: Blending what people want, what technology makes possible, and what is viable for the business to create solutions that go beyond aesthetics.

Design delivers experiences: It's about capturing insights to inspire human-centered solutions, combining products, services, spaces and information into ideas that work and people love.

Design visualizes new economies: Enabling new material cultures driven by a comprehensive design process: good for the environment, social equity, and profitable. Bringing to life new technologies and interactions. Envisioning change, creating and clarifying complex systems.

Design enables collaboration: As a process, design thinking is a catalyst for weaving together a variety of disciplines in a company to generate innovative and disruptive solutions.

Design empowers leadership: Inclusively leveraging the design process to empower people with multiple perspectives to solve problems and generate new opportunities. Bringing the organization along to communicate vision and strategy with clarity.

CREATING MEANINGFUL FOOD EXPERIENCES

It seems appropriate that this chapter should conclude with a focus on food and design. Perhaps most of those that are involved in the food industry—either from the technical or the business perspective—don't realize just how much designers love their food. Any creative session without a doubt will have good food at play. It could be said good food influences good design. And speaking of "playing with your food," how many times did most of us hear that growing up, at the dinner table? Many of us designers as kids could not help ourselves, rearranging the food on our plates, cutting our food in specific and precise shapes, or even sculpting our food, like mashed potatoes, into interesting and playful forms. It is ironic now that one of the more interesting emerging areas of design is "food design." In the last few years the emergence of television cooking shows and talk shows featuring the world's best chefs, as well as competitive game shows like Top Chef, have placed food into the same creative limelight that fashion designers and artists typically dominate.

This trend is not just about "food styling"—arranging food on plates to look as presentable and appetizing and seductive as possible. There are technologies and social movements that are converging and bringing food design into the forefront of design attention and opportunities. Design is collaborating with food scientists, nutritionists, restaurateurs, and other culinary experts, to bring innovative food experiences to the world. Technologies in food manufacturing, 3D printing, and packaging are enabling new and unique ways of engaging with our food, and how food is formed and consumed. The recent consumer consciousness, almost urgency, of knowing where our food is coming from, along with organic and slow food movements, have given rise to kitchens becoming personal laboratories of exploration, whether it is families learning how to cook healthier meals or young chefs experimenting with fusion foods and delivering unique dining experiences, creating cult-like followings. Cities like Barcelona, Portland, Oregon, and even Las Vegas have reputations as "foodie cities" and are becoming destinations for famous chefs to launch their visions. Farmers' markets and food trucks are transforming local food shopping and eating experiences into educational and cultural gatherings.

For designers, Barcelona continues to be the Mecca for food innovation, from Ferran Adrià of El bulli, the godfather of molecular gastronomy, which began 20 years ago, to today's Natural Machines, a company developing a 3D printer that will produce chocolates, pastas, breads, or anything that starts as a dough, paste, or stiff liquid. This passion for unique food experiences has spread across the globe. It seems like a daily news event that we hear of another food printing company in Europe or Asia or in the United States. Here in California, there is a rich heritage of unique dining experiences—the food hop of the 1950s, the family beach barbeque, the sushi craze and the "California Roll," outdoor dining rooms, and now the proliferation of food trucks. Chef and author, Alice Waters, the pioneer of California cuisine and the organic, locally grown food movement, has spurred a healthy food revolution and brought an appreciation of slow food to balance our fast-paced culture.

Food design is emerging as a new design and innovation sector—whether designing with or for food, food product design or even food space design. There is an International Food Design Society, with an international network of chefs, designers, academics, technologists, and enthusiasts contributing to the development of food design. There are conferences like the International Conference on Designing Food and Designing for Food, featuring Dr Morgaine Gaye, a food futurologist, or the International Symposium on Food Experience Design. Food design is recognized as a new competitive design award category in the 2013 Core77 Awards—defined as the "design at the service of eating, cooking, serving, distributing, or experiencing food. Samples include: edible materials, culinary design, delivery systems or architectures, urban gardening, educational programs, multi-sensory gastronomic experiences, kitchens, food trucks, etc."

More food-related design projects are popping up everywhere, including at Art Center. A most recent project in collaboration with TAMA Art University (Tokyo, Japan) typifies this obsession with food design. "Influencing Dining: California LifeStyle" explores what's next in dining experiences influenced by the local multi-cultural mash-up of cuisines found in Los Angeles, along with all the previously mentioned food trends. The project explored the rituals, ceremonies, and actions of the entire dining process, from the forms of the food, to the presentation of the food, to the utensils and plates, tables and chairs, and the spaces, including lighting, ambience, and mood. Food design is in its early stages as a profession and with the convergence of all of the trends, from emerging technologies to social design, the future of food design is sure to be revolutionary.

CONCLUSIONS

In Chapter 2 we discussed the expanding role of design in many areas that influence our lives, with special emphasis on the food sector. The preconception that design is simply the act of making things look pretty and that designers come into the development process only after the product has been engineered is discussed and strongly rejected. This chapter also discusses the different disciplines of design and it is emphasized that this particular chapter is written through the lens of industrial design.

- Design is more than the act of making things look pretty; it has become holistic and touches all areas of our lives.
- The best-known forms of design are graphic, advertising, fashion, interior, and architecture. There is a lot more to design and more education is needed.
- Industrial design is comprised of product, transportation, and interaction design, yet not engineering. This is often mistakenly believed and this book, and especially this chapter, is a contribution to correct this.
- Design transcends from solely object driven to both object and process driven. This is an important recognition of the big picture of this entire book, namely that designers have the capabilities to combine their striving for the ideal combination of form and function with their ability to bring new, creative, and unexpected ways of thinking to business, but also to our lives in general.
- The industry is shifting, largely driven by technology, especially using new platforms and new options such as social media, obviously the internet, and especially the recognition that advertising platforms have become multiple times more diverse. This influences ways and speed to market and calls for new retail models. Open innovation and its various forms, such as innovation partnerships, crowdsourcing, and especially the "immediacy" of information and possible solutions has substantially changed the game, and has made the economy more creative and especially more participative and more open to sharing.
- The culture of going out and doing it, the "maker culture" has taken strong roots, which is reflected by the surge of media platforms, be it print, internet, or TV.
- New technologies such as 3D printing have fundamentally changed the ways of design and its influence on industry.
- This chapter also discussed the topic of being design driven, especially through platforms such as "human-centeredness," holistic thinking, problem solving, and

partnership. We especially discussed and analyzed the concept of "weniger aber besser" (less but better), which was especially spearheaded by the well-known designer Dieter Rams.

- We then discussed the role and value of the designer in industry and society, and the new role that designers play today, largely triggered by a few iconic personalities in the world of business and design.
- We introduced and discussed the concept of the design entrepreneur and their sibling, the design intrapreneur. They both can strongly influence our lives, be it within the company or organization in which we work, but also in our day-to-day activities.
- In this chapter we discussed the impact that design, and especially the design entrepreneurs have, especially due to the fact that design has reach beyond the act of bringing ideas to life or making objects. Design has processes, techniques, tools, and skills that create impact. We analyzed topics such as: design as a process of exploration (making), design as aprocess of creating relevance (meaning), and design as a catalyst for change (transforming).
- We discussed the future of meaningful product experiences in which design played a crucial role and delivered the experience to the consumer. The following themes were briefly introduced: design creates meaningful value propositions, design delivers experiences, design visualizes new economies, design enables collaborations, and design empowers leadership.
- Finally, we discussed the role of design and designers in creating meaningful food experiences. We analyzed concepts such as "food styling," i.e. arranging food on plates to look as presentable and appetizing as possible, very well knowing that eating with pleasure can be the healthier way.We briefly discussed the well-known "food designers" and their role, and in particular the fact that food design is a recognized discipline, increasingly taught at many institutions around the world.

TOPICS FOR FURTHER DISCUSSION

- Discuss and critically analyze the role of design in your work environment. Does it play a role at all and if yes, is it adding enough value?
- Discuss with your peers, how especially industrial design in all its forms can substantially contribute to value creation in your company.
- Analyze the role of design thinking and how the combination of object driven and process driven can transform your work environment, your ways of working, your company, and/or your industry.
- Further discuss the role of design and new technologies and whether they are best used in your work environment; make proposals as to how you could imagine better ways.
- We discussed the increasing "maker culture" in today's society and industry. Try to find how much of this you can see in your own work environment and whether this is always welcomed.Develop strategies as to how you could contribute to strengthen this maker culture.
- Discuss and critically analyze the "weniger aber besser" (less but better) concept introduced by Dieter Rams and whether you find any of this in your company. Do not confound "less but better" with cost saving.

- Discuss and analyze with a group of your peers the role of design entrepreneurs versus intrapreneurs in your own company. Do you see any of this? If yes, is it effective? Can it be improved? How? If no, how can you contribute to make it happen?
- Organize a workshop in your company on the topic: Design Impact: Making, Meaning, Transforming. Discover your company's stance and commitment to design in your organization.
- Specific to a food environment, analyze and discuss the role of "food styling," especially styling that goes beyond the plate and right to the heart of the product and process.
- The authors of this book would very much welcome any feedback on this topic, as well as the outcome of your other deliberations and findings.

REFERENCES

R. Ball (2012). *Design Direct—how to start your own micro brand*, Amazon Kindle Edition.

D. Pink (2006). *A Whole New Mind: Why Right-Brainers Will Rule the Future*, Kindle Edition, Riverhead; Reprinted Updated edition.

R. Botsman and R. Rogers (2010). *What's Mine Is Yours: The Rise of Collaborative Consumption*, Harper-Collins eBooks.

A.G. Lafley and R. Charan (2008). *The Game-Changer: How You Can Drive Revenue and Profit Growth with Innovation*, Crown Business; 1st edition.

D. Norman (2007). *Emotional Design: Why We Love (or Hate) Everyday Things*, Basic Books; 1st edition.

3 How food companies use technology and design

The first rule of any technology used in a business is that automation applied to an efficient operation will magnify the efficiency. The second is that automation applied to an inefficient operation will magnify the inefficiency.

Bill Gates

FORM AND FUNCTION IN ACTION

It was mentioned and emphasized repeatedly in Chapter 1 that the best design results from a balanced combination of form and function. As nature demonstrates time and again, a tree can only grow high, i.e. follow a specific pattern of design, if this goes hand in hand with the specific and deliberate functionality of creating strong and long enough strands of ligneous and cellulosic material with the optimal build and strength to support the growth of the tree. And it does not only support the tree to stand there, it also supports the tree under stressful conditions such as heavy wind, rainfall, or loads of snow on its branches.

Design that is crafted by humans, by designers and people who work in this field has to follow the same pathway and this means that designers have to gain and apply engineering knowledge: this is where technology really comes into play. It is this combination and interplay between design and technology, i.e. between form and function, that makes for successful and sustainable, meaningful design.

Some of the oldest examples of a great combination between form and function date to ancient times, approximately 450 BC, and can be admired in the Ancient Agora Museum in Athens. Figure 3.1 shows two antique grills; their shapes are truly familiar and timeless, simply because they are built on the concept of form and function.

Figure 3.2, even more striking and almost modern, shows an ancient BBQ, also from around 450 BC, and also to be found in the Ancient Agora Museum in Athens.

Great designers of all periods truthfully follow this mantra, as can be seen in the works of fifteenth century genius craftsman and artist Leonardo da Vinci. Figures 3.3 and 3.4 show both the genius of creativity and invention, and a deep technical, even engineering understanding of the concepts and ideas proposed by da Vinci in those days.

Food Industry Design, Technology and Innovation, First Edition.
Helmut Traitler, Birgit Coleman and Karen Hofmann.
© 2015 John Wiley & Sons, Inc. Published 2015 by John Wiley & Sons, Inc.

Figure 3.1 Portable grills; Ancient Agora Museum, Athens, Greece (Copyright: Hellenic Ministry of Culture and Sports/Archeological Receipts Fund. Reproduced with kind permission of the Hellenic Ministry of Culture and Sports.)

Figure 3.2 Terracotta oven; Ancient Agora Museum, Athens, Greece (Copyright: Hellenic Ministry of Culture and Sports/Archeological Receipts Fund. Reproduced with kind permission of the Hellenic Ministry of Culture and Sports.)

Successful modern day designers follow the same intuition and drive; namely a deep understanding of the why and what, you may call it the "inner life," the interior of every object they are supposed to design in an optimal way. This deep understanding, and only this, is a guarantee of successful and sustainable design, or else it will remain a hollow shell.

Let us briefly talk about Jonathan Ive and his groundbreaking designs for the Apple company, more specifically about his approach to design in general. "Ease and simplicity of use" are Ive's watchwords, principally his mantra. His designs show a deep understanding of technical needs (the function) and the resulting product shapes (form), both of these being in perfect harmony.

Figure 3.3 Leonardo da Vinci's sketch of a flying machine. (Reproduced with permission of Google.)

Function in good design has not just to be in harmony with form, but has to be fully understood by the designer. This is the next step up from simple balance, namely to intimate understanding, to potentially challenge the technical needs and requirements of a product, and, if necessary for good design, even to tweak the technicalities so that they fit the most appropriate design. A good example of this is the reshaping of the processor in the 1998 iMac so that the very colorful, almost candy-shell could be executed in the well-known, iconic fashion. Ive's design for the 2000 Power Mac G4 Cube was executed in such a way that the cube could easily be removed from its housing so that the user had access to the interior of the computer and free air flow was enabled, thus allowing for fan-free operation, with a much lower annoying noise level. Eventually, desktop computer and display screen were integrated into one unit, which resulted in the notorious flat-panel iMac of 2002 and the following generations to this day.

Although already over 10 years old, these designs not only show a kind of timelessness, but almost didactically lead towards the designs of today, obviously not only from Apple. However, there is one more example where design, through a very deep and intimate understanding of the underlying technology, pushed the bounderies. The Mac Air consistently used the concept of the stationary hard-drive, i.e. no hard drive at all, but just a very large static storage device, which enabled this next step of miniaturizing and making laptop computers even easier to carry and use on the go, as a full fledged computer with processor, display, storage, and keyboard, all only a few millimeters thick.

Figure 3.4 Air-screw drawn by Leonardo da Vinci. This idea presents the basis for today's helicopter. (Reproduced with permission of Google.)

We most certainly can find many more examples of designers that have a deep understanding of the technical aspects of the product that they are about to design and who are willing and capable of applying this. However, few have gone as far as Jonathan Ive did in his designs for the Apple company.

IMPORTANCE OF DESIGN IN THE CONSUMER GOODS INDUSTRY

Let us take a closer look at the fast-moving consumer goods industry and how consumer goods companies and food companies have applied this approach of combining deep technical understanding of their product with optimum, inclusive, and universal design. To better understand this, we need to look in more detail into the ways that such companies use design and designers within their own organization, but also by collaborating with external resources in various partnerships.

As a first example, we will look at Proctor & Gamble (P&G) in some detail. When P&G came under the new leadership of Alan G. Lafley, over 10 years ago, there was a clear recognition that, because the company was in some ways underperforming, new directions needed to be taken in many areas, one of which was their approach to design and how design should best be used. At around the same time, P&G started their very much talked about and often copied open-innovation journey under the name "Connect & Develop" (C&D).

Open innovation is an approach in which mostly technical organizations (companies, universities, and others) recognize that they do not have all the expertise they need in a specific area and therefore ask outside sources. Many ways of open innovation are conceivable, and many have been tried. For more information, Henry Chesbrough's book, *Open Innovation*, is one of the best sources (Chesbrough, 2003).

Let us critically review both of these events, namely the decision to approach design in new and different ways and at the same time to look for creative innovation and resources in the world outside the walls of P&G. In the years between 2002 and 2005, P&G hired an estimated 200 designers, many or most of them industrial designers, and started a very strong design program under the leadership of one person, Claudia Kotchka. She reported directly to the CEO of the company, but was rather far detached from board level. As much as this setup was the envy of many competitors in the industry, reality probably looked slightly different from the inside, given the number of newly hired designers competing for interesting and exciting projects with a chance of winning through to the very end of a project and into consumers' homes.

A critical view may be expressed here, namely that whilst P&G initiated the widely recognized and heavily marketed C&D program that should and could have found numerous talents in the design world outside P&G's company perimeter, they hired this large number of designers into the company. In hindsight, it looks like this did not make as much sense as it may have appeared to then, but it made for excellent publicity in the professional community, as well as with the public at large. Many articles were written that referred to this topic and many speeches were given by the protagonists themselves (Kotchka, 2008) as well as "third parties" (Martin, 2009).

As over-reported and over-exploited as this approach by P&G may possibly have been, and as often as it was cited as the one and only example of an FMCG company approach to design and how much design and innovation was embraced, it had a definitive provocative effect on other, similar companies in the area of consumer goods, especially food.

We will discuss P&G's open-innovation approach in more detail in Chapter 11, and especially in comparison to other, similar yet simpler and equally efficient approaches. In this chapter we will focus on design and how P&G's total embrace of industrial design affected their competitors and shaped the landscape of design in large parts of the food industry over the last 10 years. We will largely use the example of Nestlé S.A., the world's largest food company, to illustrate the use of and approach to design in the food industry in general, and to compare it to P&G's design approach.

When we measure both approaches against the main paradigm of this book, namely to establish design thinking and designers at the executive board level of a company, especially a food company, we can simply answer that although they are both "light years" away from this situation, some good and interesting baby steps may have been taken already.

Roger Martin's book on "design thinking" insinuates that there should be more design thinking in corporations and he largely quotes the P&G example mentioned a few paragraphs above. "Design thinking" sounds like the real thing (real meaning coming pretty close to our stated paradigm), but it does not go the extra step. It describes the importance of design, but somehow neglects the importance of designers and their capabilities to influence the strategic directions of a company, especially a food company. This is exactly what we intend to suggest in this book, and to present and discuss examples that support this.

Let us get back to the real playground of design (graphics, industrial, media, and communication) in the food industry, how this is typically perceived and the typical consequences.

When I first became exposed to packaging at the Nestlé Company (USA and then global) in 2003, I was told the following about packaging (first the "good-speak" and then, in brackets, what it really meant):

- Packaging cuts too much into our profits (It's too expensive, find cheaper materials)
- Our products are over-packaged (Again, too expensive, make packaging with less material)
- Do we really need this type of packaging? (Get rid of it as much as possible …)

- Did you see the great-looking product from the competition? (Our packaging sucks, add another color ...)
- We should re-organize packaging, our approach to it, as well as the people involved (Get rid of it)
- Marketing plans a re-launch (Make the packaging cheaper still)
- Etc. etc. Note: Reader, please find your own comments and kindly share them with us!

In short: packaging was rather seen as a nuisance, too expensive, not strong enough in its expression to the consumer, and ideally should have been reduced as much as possible. In those days, around 2003, there was a total focus on the material and cost of packaging, which, certainly thanks to the design movement at P&G, was the starting point of a great ride to change things, quite a bit, actually quite a lot! Design was not necessarily at the forefront of the thought process of how packaging could and should be approached at the heart of the Nestlé Company, but we went back to the drawing board, almost a blank sheet of paper.

Reader, please go back to Chapter 1 and have a quick re-read of the paragraphs on "inclusive design"; this notion of inclusive design became the ever present driver for everything that was discussed, deliberated, brain-stormed, re-discussed, proposed, planned, and ultimately decided upon within the company. On the basis of this, we defined a concept for packaging under the heading "The Seven Drivers of Packaging Innovation and Renovation".

THE ROLE OF TECHNOLOGY AND DESIGN IN PACKAGING INNOVATION AND RENOVATION

Figure 3.5 depicts an overview of this concept, and in the following we will discuss the individual points in more detail and how these are linked to design in a food company such as Nestlé (H. Traitler, personal communication, 2004).

The important message that comes from Figure 3.5 is to realize that we always deal with two stakeholders, i.e. the very people, inside as well as outside the company, who are affected by every action and decision taken in the context of packaging innovation and renovation, as well as the technological necessities.

Food safety, quality & environment

Packaging sealing quality: integrity of the package

Let us begin with the area of consumer relevance and, as already discussed in Chapter 1, food safety, quality, and environment are the foremost concern. When it comes to packaging, from a technical standpoint, features such as better sealing quality lead to a better protection and keeping quality of the food product. Improved sealing quality of, for example, flow wraps for confectionery bars, is achieved by two major directions. The first lies in the appropriate sealing technology, meaning that the right tools, i.e. the correct seal jaw designs, are used, the sealer is correctly set in efficiency, temperature (except for cold seals), and speed, and that as few micro-fissures as possible exist in the seal. The second

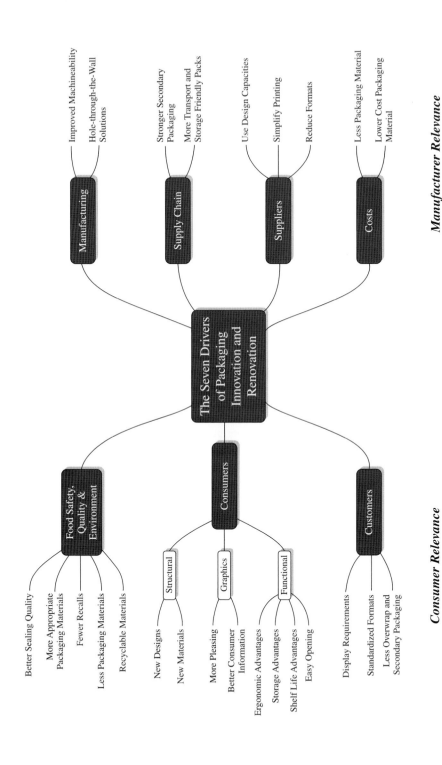

Figure 3.5 The seven drivers of packaging innovation and renovation in the food industry. (H. Traitler, Nestlé / Nestec Ltd., personal communication, 2004.)

is that it is important that the right, bespoke packaging material with the relevant sealing layout—cold seal or hot seal—is chosen that can be properly sealed by the sealer, leading to a quasi-perfect seal.

These two aspects are not trivial because the entire operation is always under cost pressure and the perfect balance between material choice—sealing efficiency—line speed is not necessarily always achieved by the lowest cost materials at the fastest line speed. Depending on expected shelf life and accepted olfactory quality losses over time, compromises can be found and applied. As briefly mentioned in Chapter 1, typical supply chain transit times for a confectionery product, i.e. the time it takes from the day of production to the mouth of the consumer is four months on average, and the sealing quality of a good package is normally optimized to that standard, although it would be an over-statement to say that achieving this balance is an exact science. Important strides have been made from a technological as well as an olfactory point of view to find the most efficient and cost-efficient compromise.

More appropriate packaging materials

It might go without saying that the industry always uses the most appropriate packaging materials for the specific job at hand. But is that always the case, or do cost considerations sometimes over-ride this almost natural paradigm? From rather detailed personal experience I would say that in almost all cases appropriate materials are used, but then, all materials are always undergoing development, improvements can and do happen, and the newly improved, more appropriate materials are not always used immediately, for reasons of habit or standing supply contracts. Let us talk in some detail about "appropriateness" of materials. In general we have four over-arching themes to consider:

- Barrier properties for oxygen, water, and light (for flexible and hard plastics, aluminum, glass, metal, and composite materials)
- Mechanical properties, i.e. strength, puncture resistance, breakability, folding behavior (for all of the above, paper, and carton/cardboard)
- Costs (for all of the above)
- Environmental performance.

There is one more theme, however, as especially recyclability is of increasing importance, but also potential re-usage is considered an important aspect by the industry, as well as by customers and ultimately the end-consumers.

Especially when it comes to liquid products in plastic packaging, especially bottles, there is a strong trend within the industry, as well as with consumers to use more and more PET (polyethyleneterephthalate) bottles. These bottles have been used for a very long time in the soft drinks industry, but not so much in dairy applications, for instance. PET has rather limited but satisfactory water barrier properties, excellent oxygen barrier properties, but obviously no light barrier, unless one glues a dark label all around the bottle.

If the manufacturer wants to showcase the product inside, this is not the best way to go. More recently, even aseptically filled dairy-based products have successfully been packaged in fairly transparent PET. The culprit with light is not so much the UV region—there are transparent UV barrier solutions offered by suppliers—but rather the visible wavelengths. It is almost impossible to find a chemical that can be added to PET that will absorb visible light and still leave the bottle wall transparent. Many efforts are underway in the industry, on the supplier side as well as the food manufacturer side, to find acceptable solutions. (Ch. Saclier, personal communication, Nestec Ltd., 2009).

In the hard plastics area, PET has pretty much taken over the leading role for food packaging, although HDPE (high-density polyethylene) and PP (polypropylene) are still in the running. Polycarbonate, despite its great mechanical strength and thermal properties, has come under rather critical scrutiny, especially because of the danger of, admittedly miniscule, amounts of the residual monomer BPA (bisphenol-A). In many food areas it is therefore increasingly replaced by the above-mentioned packaging materials.

As far as flexible plastic materials are concerned, PE (polyethylene) is still a strong contender, increasingly being taken over by composite materials in combination with aluminum. More recently, mainly for recycling reasons, aluminum is increasingly replaced by combinations of different PET materials and PE in duplex or triplex arrangements. We will revert to some of the above considerations in more detail when we discuss the area of "manufacturer relevance", and especially "supply chain" and "suppliers."

In conclusion, it may be said that consumers want to see "honest" packaging, i.e. packaging that gives the feeling of being just right, in amount as well as protection, is easy to discard or can easily be recycled, and, interestingly enough, especially in the case of flow wrap or wrapping materials in the confectionery industry, does not have a memory effect once the consumer has eaten the sweet and crumples the package to a small, ideally tiny little ball of close to nothing … From personal consumer observations and consumer tests, a feeling of guilt is directly linked to the volume of residual packaging material lying in front of the consumer: the higher the apparent volume, the higher the guilt, the smaller the crumpled packaging residue, the smaller the guilt! We call it the "crumple-guilt" complex.

Less packaging material and recyclable materials: the three Rs, recycle—reuse—reduce

This is a rather obvious topic, as it has a positive connotation for all three concerned parties: the consumer, the customer, and the manufacturer; all of them wish for a world with less packaging material, by simply reducing the amount of packaging materials used in the food industry. It is a myth that food manufacturers want to use more packaging material than necessary, just to make their product look bigger, more important, or more impressive. There are certainly exceptions to be found in the marketplace, but, from personal experience and numerous visits in retail shops and supermarkets in all parts of the world, they are rather rare.

The prominent exception in the consumer goods industry is the cosmetics and fragrances industry, which definitely has a tendency to over-package and use more packaging material than necessary, however in the food industry this tendency is rather absent and the opposite may be the case. Again, we will discuss the role of primary packaging (the one that goes around the product) and secondary packaging (the one that helps the product transition through the supply chain) in more detail later.

With increasing public pressure to reduce waste material, the individual, the consumer, has a strong desire to buy products with as little as possible packaging material and to be able to simply and cleanly separate and recycle them as much as possible. This is a very real scenario in many developed countries, Switzerland showing one of the highest recycling

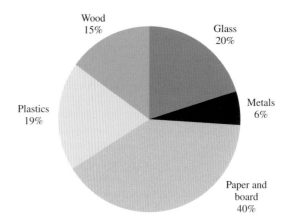

Figure 3.6   Percentages of packaging waste by weight, EU-27, 2010. (From: Eurostat.)

rates for glass bottles, aluminum cans, and PET of above 90%, with an average recycling rate, i.e. over all recycled materials, of over 75%, the latter being the target rate for the Swiss government (Cleantech-Switzerland, 2012). Other developed countries' typical recycling rates are as follows:

- The European average target for recycling of all "packaging waste generated and recycled … all packaging materials including glass, 'paper and board', metals, plastics, others and wood" was 55% in 2008 (European Commission, 2013).
- Almost all European countries have achieved or slightly over-achieved the target of 55%.
- When looking at PET alone, recycling rates in Europe are, on average, 51% and in the USA a rather disappointing 29% (PET Resin Association, 2012).

Figure 3.6 shows the proportions of different packaging wastes in Europe, as reported by Eurostat for 2010.
In the USA the situation is as follows:

- The rate of recycling of municipal solid waste (MSW) in the USA was at an overall 34% in 2010, with PET accounting for only 29%, i.e. still a long way to go, especially for PET.
- Glass bottles and containers were recycled at a rate of 33%, probably helped by the fact that a small amount of money can be made by returning bottles and also cans (EPA, 2012)
- Recyclable packaging materials are increasingly requested by consumers, but also by the industry, which relies more and more on recycled materials in all areas, cardboard and plastics, as well as metal and aluminum containers, for their operations.
- Typically the soft drinks industry counts on an average of one-third recycled PET for making new plastic bottles.
- The Coca Cola Company, as just one example, was very much at the leading edge of recycling of their PET beverage bottles and uses the argument of recycling and

therefore helping the environment as a message in commercials and publications (The Coca Cola Company, 2014).

- Finally, re-using packaging materials such as glass bottles or PET containers without having to re-work them has limited potential, especially in the food industry, simply for hygienic and potential toxicity reasons as it is not always clear for which purposes such containers were used before they might be returned for re-usage.
- A total re-work of the materials, glass or plastic, is therefore the safer and preferred way in the industry.

Fewer recalls

This is a very straightforward and much desired feature by both consumers and manufacturers; it simply signifies that the lower the number of recalls, and here we are talking about recalls due to technical, mechanical reasons and not wrong formulations or wrongly declared ingredients, the more efficient the supply chain, the lower the extra costs for the food industry and the higher the confidence of the consumer in their product of choice. Reduction of recalls can be achieved through two main pathways:

- Most appropriate packaging material selection
- Most efficient storage and distribution in the supply chain.

We will discuss both of these aspects in more detail in the following section.

Supply chain

As briefly mentioned above, manufacturers have a strong desire to optimize both primary and secondary packaging, especially for reasons of durability through the outbound supply chain, i.e. warehousing and distribution. Of special importance here is stronger secondary packaging, which means stronger, more adapted cardboard, better design, and more clever structural orientations of cutting such secondary containers in order to achieve improved structural strength.

More is not always better, but some basics need to simply be respected. For instance, 24 KitKat Chunky candy bars in a carton have a weight of approx. 1.5 kg, including the carton, give or take.

Arranging them on a pallet for distribution may mean that there might be between five and as many as ten cartons stacked on top of the other, resulting in a weight of several kg, maybe up to 10 kg sitting on the lowest carton. Because there is no differentiation between cartons and no one carton is different from the others, all secondary containers have to be able to hold up such weight, not just for a short period of time, but sometimes up to several weeks if not months. Not only weight plays a role, but of equal, if not greater, importance are storage conditions, such as humidity and temperature.

Whilst temperature is often much more critical with respect to the keeping quality of a food product, it is humidity that takes the greatest toll on the stability and integrity of secondary packaging. On the other hand, when it comes to protection of powdered packaged products in the supply chain, it is all about mechanical protection.

The material supplying industry goes to great length to come up with even better cartons, more moisture resistant, better mechanical strength and less volume, but it almost sounds like squaring the circle and the title of a very funny and successful Hollywood movie: *Something's got to Give*. Most frequently it's the volume, the dimensions of secondary packaging that simply has to be up to the task, irrespective of the general desire for having less.

Optimal pallet loading is another important aspect to optimize the supply chain. Many products, whenever possible, are retro-engineered in their dimension so that a multitude of single products in a secondary transport and distribution container can be arranged in simple ways to optimally fill the surface of a pallet with neither "under-hang", i.e. too much empty surface on the pallet, nor "over-hang", thus not allowing pallets to be closely arranged in warehouses or distribution trucks. From personal experience, an over-hang of 1 to 1.5 cm is largely preferred to under-hang.

Overall, the goal is to have packaging solutions that allow for clean, safe, mechanically sufficient, and easy to pack and unpack secondary transit packages, as well as individual product packages. The customer (retailer) requirements will briefly be discussed in a later section (Customers).

Suppliers

The role of suppliers, especially packaging and packaging materials suppliers, is still underestimated in the food industry. However, it can be said with absolute certainty from long-standing personal experience that the situation is even worse if not to say much, much worse with the collaboration between ingredient and technology suppliers, at least as far as the food industry is concerned. Suppliers are a great source of new ideas and innovative solutions; maybe not all of them, but certainly most of them.

Again, from personal experience it can be said the there is a negative correlation between the size of a supplier company and the degree and frequency of innovation, and clever and applicable innovative solutions, especially in the packaging world: the smaller the supplier company, the higher their desire and, even more so, need to be perceived as very innovative and flexible by the large customer called the food industry. The problem, however, is that these sizes very often do not match, and smart approaches have to be found and worked out in order to achieve the best possible solution for both parties. This can be done by appropriate collaboration agreements, temporary exchange of development experts, and similar activities. At the end it's all about commitment from both partners to go to the very end of a creative packaging development and being able to manufacture the resulting solution in adequate quantities in the desired timeframe and geography, i.e. in the location it is needed.

Costs

We tend to believe that innovation in any domain can be achieved at cost neutrality or even lower costs.

This is certainly true when it comes to game-changing breakthrough solutions in any type of industry, but is often not achievable in more trivial, transitional types of innovation. Packaging is such an area, where innovation happens in small steps, is transitional and not transformational. I prefer this terminology over the widely used definitions of "incremental" and "radical" innovation. These two types of innovation are often also linked to "consumer pull" for incremental and "technology push" for radical innovation.

From personal experience, it is rather doubtful whether such an easy and straightforward link can be made, but for the moment we can accept this distinction. We will discuss innovation in all its complexity and consequences in later chapters, especially Chapter 8.

Now, what's the link to costs? Very simply, innovation has a cost and innovation at zero cost is almost a dream, often dreamt by marketing people, who, quite understandably, have very little or no knowledge of the complexity and inherent difficulties of technical innovation. However, as an example, optimization of sealing quality with subsequent improved product keeping and taste qualities can result in cost benefits. Yet, in the "real world" of corporations and business there needs to be a connector, and such a connector exists: it's the designer and the design-thinking process, unfortunately today still very under-utilized, if recognized at all.

The reality today is that many innovative solutions, especially in the world of packaging, but also in product development, never see the light of the day because they are simply considered too expensive.

It must, however, also be said that technical developments are often complex and take too much time, thus once they are ready to apply, at whatever cost, the market may have moved on, and what was exciting and relevant two or three years ago is no longer of interest. Sometimes this may sound like a marketing person's excuse, but more often than not, unfortunately, it reflects the reality.

There are three possible solutions to this dilemma:

- Be faster in the development phase
- Be especially faster in the validation phase (which can sometimes drag on the longest!)
- Use additional external resources to speed up the process and potentially reduce costs.

We will discuss these aspects in more detail in Chapter 11. For now, suffice it to say that most of the more trivial, transitional innovations, especially in the packaging area are centered around graphics adaptations or using new (or new in this particular application) packaging material which is hopefully cost neutral.

The following anecdotic observation was made in the context of costs and color printing. When discussing with the printer converter to add color number 6 to the five already existing on, for example, a candy wrapper to make it (hopefully) a bit more attractive, a certain additional cost is to be expected and such a premium is typically accepted and paid. When talking to the same printer converter, asking to reduce the number of colors on a candy wrapper from, for example from six to five to make the look sleeker or less complex, the standard answer will be: yes, we can do, cost-wise it's no big deal, it'll be roughly the same cost. Sounds familiar? I have experienced this time and again. Anyhow, let's not complain, but rather say that tough negotiations with a deep understanding of all aspects, design, marketing, and underlying technologies, definitely helps to achieve the most appropriate and acceptable costs for a packaging or product innovation in the food industry.

Consumers

Everyone in the industry understands this and must have heard it a thousand times: "The consumer is at the heart of our operations", "The consumer is king (or queen for that matter)," and, of course, it is very true; we must never forget it! When it comes to consumers, proper design of the packaging solution is of utmost importance. This is the area where design and technology meet each other in important ways and the following are important considerations.

Structural aspects

There are two main considerations to observe here. The first centers around new designs, really new and surprising, but most of all functionally logical shapes and expressions of the product through great packaging design. In a way, packaging "is the expression of the soul of a product" (P. Brabeck, personal communication, 2004) and the design of such expression of the soul is of critical importance, not only in structural, three-dimensional ways, but also by the most appropriate graphic expression. The latter will be discussed in more detail in the section on "Graphics."

When we talk about "the soul of a product," it insinuates that we know the soul of the product, or in other words that we have a very clear and intimate knowledge of what the particular product stands for and how the brand essence is defined. This is not at all trivial and many new generations of junior and senior marketing people have scratched their heads around the question: "What is the essence of the brand that I am marketing?"

Figure 3.7   Nestlé Carnation liquid and powder Coffee-mate®. (Reproduced with kind permission of Nestlé SA, Vevey, Switzerland. Nestlé, Carnation and Coffe-mate are registered trademarks of Société des Produits Nestlé S.A., Vevey, Switzerland.)

Figure 3.8 Nestlé Coffee-mate® prior to the packaging change. (Reproduced with kind permission of Nestlé SA, Vevey, Switzerland. Nestlé, Carnation and Coffe-mate are registered trademarks of Société des Produits Nestlé S.A., Vevey, Switzerland.)

Until this question is not answered in a consistent and notoriety-building way, no good and appropriate packaging can be designed.

New designs can often be achieved by introducing new materials that allow structures and mechanical properties that were not achievable before. Packaging designers often see this as a great opportunity to introduce new, even groundbreaking, packages, as was very impressively demonstrated by the introduction of the then new containers for both powder as well as liquid Nestlé Carnation Coffee-mate® back in 2001.

The images in Figure 3.7 show this transformation. The new material was plastic, going away from glass and thereby allowing new functionalities, as well as shapes with great haptic, but also grip performances.

The image in Figure 3.8 shows the old glass container in which the product was filled and sold prior to the packaging change.

Graphics

It's not so long ago that the aspects of graphics and two-dimensional appearance were the only ones recognized as design by many in the food industry. When asking colleagues at all levels what they understood by design, the answers would be: a new script font, a change in color, an additional color, more written information on the pack, or a new or modified brand logo. Other features were maybe recognized but this was not design, this was "getting the package right, especially making it less expensive."

With the increasing introduction of plastic materials of all possible types since the early 2000s, it was recognized that good packaging, as well as product design, goes beyond just graphics and it is now well understood that structural ("3D") and graphic ("2D") design go hand in hand.

In focusing on graphics design, from all our practical work, two features clearly stand out, and in a way they are almost generic:

- The appearance of the package has to become more pleasing, has to achieve a higher acceptance by the consumers, and has to touch emotions and "strike a chord" with the consumers.

    It is of utmost importance that this is not simply attempted by the technical packaging person ("… let's add another color, the print machine still has empty print heads … ") or by the marketing person ("… if the logo was larger, it would be more striking … ") or the designer ("… if we were to introduce this new fancy font, the harmony of my design would stand out so much better … "), but in close collaboration with consumers and with consumer understanding in mind. This is not a trivial task and requires intensive collaboration between all parties and respect for the input from each side, as well as the guts to ultimately make any change or modification.
- The second important feature of good graphic design on food packages touches the area of consumer information. Every reader can probably relate to the lengthy and, most of all, tiny print that is to be found on almost all packaged foods and beverages. There is a fine line between not saying anything at all (which in today's legal and regulatory environment is impossible) and packing tons of information, in addition to the legal requirement, on the package. Companies want to distinguish themselves by personalized information delivery and this often leads to rather busy and crammed packaging. A few years ago the Nestlé Company introduced the "Nutritional Compass," which is definitely a very useful feature for health-conscious consumers, but requires quite some space or "real estate," as it is sometimes called, on the package. Figure 3.9 shows a typical example of such a "Nutritional Compass."

Great graphic design is of utmost importance for the success of a food product and it is important to recognize that graphic design combines the eye, the brain, and our emotions in important ways and has to ideally go hand in hand with the more touchable and functional features of a good food package.

Functional packaging aspects

Functional aspects of packaging are important to all players in the value chain, from manufacturer to distributor to marketing, sales, trade, and last, but not least, to all consumers. Let us begin with the one feature that everyone asks for, but does not always find, namely *easy opening*.

A package that is easy to open should be the most normal and expected feature of a good food pack, but as many readers must have experienced themselves, this is not the case. We can speculate about the reasons, yet two stick out: first, "it's the costs, stupid" and second, it is just simple laziness of mind or resistance to change anything on my existing packaging line, even the slightest modification that would not add cost at all. It is very difficult to go against either of these two culprits and one could argue that the true low-calorie food

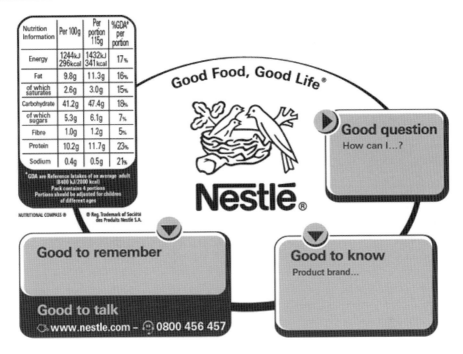

Nutrition Information	Per 100g	Per portion 115g	%GDA* per portion
Energy	1244kJ 296kcal	1432kJ 341kcal	17%
Fat	9.8g	11.3g	16%
of which saturates	2.6g	3.0g	15%
Carbohydrate	41.2g	47.4g	18%
of which sugars	5.3g	6.1g	7%
Fibre	1.0g	1.2g	5%
Protein	10.2g	11.7g	23%
Sodium	0.4g	0.5g	21%

*GDA are Reference Intakes of an average adult
(8400 kJ/2000 kcal)
Pack contains 4 portions
Portions should be adjusted for children
of different ages

NUTRITIONAL COMPASS ®   ® Reg. Trademark of Société des Produits Nestlé S.A.

Good Food, Good Life*

Good question
How can I...?

Good to remember

Good to know
Product brand...

Good to talk
www.nestle.com – 0800 456 457

Figure 3.9 The Nutritional Compass. (From: The Nestlé Company, http://www.nestle.com/brands /nutritionalcompass. Reproduced with kind permission of Nestlé SA, Vevey, Switzerland.)

is the one that stays in its package because I cannot open it. Just a joke, of course, but from personal video-ethnographic observations (video ethnography being the technique of filming consumers' actions in their own or a standardized environment in a kind of "candid camera" situation and subsequent observation and deduction), I can say that I have seen consumers giving up attempting to open a package and putting the product aside. It is also safe to say that this was probably the last time that this consumer would buy this product. Therefore it can be deduced that one very difficult-to-open package equals one or several lost sales. Getting the opening right and making it easy-opening for consumers is probably the single most important feature in the creation of great and attractive food packaging.

Failing in this means difficult times ahead for product sales.

Now, we have mentioned the excuse of cost, and it is true that sometimes a good easy-opening feature may cost a fraction more. Real street-smart designers and technologists would prepare the cost case, particularly factoring in potential lost sales and subsequent potential reduced profitability, and sit together with marketing and sales and hash this out. What also helps internally within the organization is to highlight such important aspects by making them focus points and, for instance, annual targets. Such focus and target setting can then easily become a part of the objective setting for the individuals involved, thus giving them an extra supportive boost to follow such an important target and find the most appropriate solutions.

Another important functional packaging consideration lies in the *ergonomic features* of the package. These are again important for the supply chain, but also for the consumer at the end of the value chain.

Specific shapes might be of great advantage to the consumer to better grasp and hold on to a beverage container, for instance, but the same container's shape might be a real nightmare for stackability (how containers can be stacked on top of each other in safe and surface-optimizing ways), palletization, storage and transportation. Good structural design can bring this together and needs to work, again, hand in hand with the packaging technologists and logistics experts. This is not always easy, as these people might be geographically rather detached, but, from personal experience, this is the only way to achieve workable solutions.

Such solutions quite naturally lead us into the aspect of *storage advantages*. The main driver here is the recognition that good packaging functionality combines the requirements of all members of the value chain, storage being probably the one that takes the largest chunk, from a time perspective. It may take an hour, give or take, to produce a candy bar, including wrapping, secondary packaging, and palletization. The pallet may then sit in the manufacturer's outbound warehouse for a short period of time, maybe a few hours or a day, before it is put on a truck for delivery to a distribution center or directly to the retailer's warehouse, and then onto the retailer's shelf. As mentioned before, the average transit time from the day of production into the consumers' mouths is between four and five months for a candy bar, depending on the size of the market. We can easily see that in the entire chain of events, the product spends the longest time "siting around," nicely embedded in its package, waiting for its final destination, namely to be eaten. Storage advantages that combine necessary and optimal features to best survive transit and waiting, with consumers' quests for handy, easy-to-open, easy-to-grasp packages can be designed by good package designers, again in collaboration with the technical experts and the logistic experts, not forgetting marketing and sales.

The last point under this topic is *shelf-life advantages*. Such advantages are largely linked to three major considerations:

- Most appropriate selection of materials. Again, cost and reduction of cost, or at least cost-neutrality are a big driver, almost a tidal wave that designers and technologists will find themselves up against. The same street-smartness is required by being prepared to factor in lost sales and especially bad-goods return numbers in such discussions and how a 5% increase in packaging material costs, for instance, can lead to a reduction of 1% in bad goods and returns, by no means an unrealistic assumption.
- Optimally functioning and well-serviced and run packaging machines and lines. This point speaks for itself; maybe just to add that continuous training of and increasingly handing responsibility to involved personnel are excellent ways to improve the situation, and again, it does pay off!
- Optimized outbound logistics, especially storage conditions during transportation and in the warehouse. This is especially a task in which the technical packaging experts and the logistics experts need to closely collaborate. The main message here is to the technical packaging group: do not lose interest ("my job is done") once the pallet leaves the factory warehouse. Continue a very close dialogue with your logistics colleagues!

Customers

To be specific, even if for the typical reader of this book it is rather clear: in the food industry, customers are retailers of any kind, small or large, the ones that sell to other retailers, as well as the ones who sell directly to consumers. Retailers have very specific needs and requirements that even sometimes direct the industry in rather strong ways, such as the "strict order" given by Walmart in the mid-2000s that their top 100 suppliers should increasingly introduce the still rather new technology of radio frequency identification tags (RFID) on at least pallet and possibly case level, in passive mode, which meant that detectors would have to be installed in the packaging lines in the immediate vicinity of the pallet or secondary carton on the pallet. Many strides were made in this direction, in which I was personally involved. Many trials and tests were conducted to find out about the field of applications, such as solid products versus liquid products, and whether detectors would still reliably read RFID tags on boxes that were not sitting on the perimeter of the pallet, but rather inside, shielded by other boxes.

This went on for a few years and, due to the fact that the benefits were not really clear at that time, the whole drive eventually dropped to a very slow pace. We have a good 25 and more years of experience with bar codes in the industry. It was also clear that, should RFID technology flourish, which admittedly could optimize inbound, but especially outbound, logistics (seen from the manufacturer's angle), over existing bar code technology, both technologies would have to be used in parallel, at least for a period of several years.

Today, it has become rather silent around RFID, as the technology to manufacture RFID chips has still not reached the "1 cent per tag" cost target, which would only be achievable by improved printing technologies so that the embedded chips could be entirely printed and would not have to be manufactured in the traditional way. Moreover, the ultimate in RFID technology would be to have "active" tags, i.e. tags that consist of the chip, an antenna, and a battery, thus, for instance, allowing detectability of products on pallets that sit in the rear corner of a warehouse some 50 or more meters away.

However, optimizing food and beverage packages so that they perform better in the supply chain and especially at the back end in the retailers' space is of utmost importance and, in an ideal world, happens with close collaboration between the retailer and the manufacturer. The following topics should be considered:

- Retailers have very specific display requirements, which need to be considered from the very onset of product and packaging design. Such displays can be brand- and product-specific "side-kicks," i.e. the displays that one typically finds at the end of an aisle. They can present a specific brand or mixed brands fitting display for seasonal or promotional occasions, and they need to ideally fit into the space between the aisles. All these requirements need to be communicated early on, so that product developers, packaging technologists, and designers can take these into consideration.
- Another important aspect is the retailers' need for more and more standardized packaging sizes that fit into their shelf space, both in single rows and stacked one above another. There is nothing worse than for a product to be displayed in an un-elegant way, for instance two rows one on top of each other and in the third row, because the package is too large, the product needs to be displayed sideways. Moreover, retailers rely increasingly on "shelf-ready" secondary packages, for instance multi-product display boxes, which can easily be opened, either on the top or the side, and put on the shelf without further handling, thus also minimizing the amount of packaging material that has to be discarded or, ideally, recycled. Again, shelf readiness should be built from scratch by close collaboration between manufacturer and retailer.
- Lastly, retailers do not appreciate receiving pallets into their warehouses or retail space that are over-packaged or, in particular, over-wrapped with shrink films. On the one

hand, such over-wrapping makes for improved transportation efficiency, on the other hand, retailer personnel need to unwrap such pallets, mostly by using very sharp cutters. This is an area where safety concerns are highest, and we speak about the "health and safety" of personnel involved in such activities.

To paint an extreme picture, too much shrink film requires substantial use of cutters, potentially leading to the operator being hurt, which needs to be reported and yields a higher number in the company's report card on accidents, making the health and safety personnel responsible very unhappy and perhaps shedding a bad light on the retailer's public standing. It's like the proverbial butterfly wing flapping in one region, causing terrible winds and thunderstorms far away.

Too much secondary packaging, initially used to improve the product security in distribution, is seen as very negative by the retailers, as they have to get rid of it, which is a logistical and cost challenge. Ultimately, it's the consumer who pays for it, one way or another!

Manufacturing

The last aspect that we will discuss in this chapter concerns packaging-related considerations linked to manufacturing. Manufacturers, especially those who operate in the manufacturing of products and packages, have a very simple mantra: less complex and less expensive. One important aspect is therefore to work with products and packaging material combinations that improve "machinability" in two ways: potentially increasing the line speed, based on the most appropriate material choice, as well as simplicity of creating and/or erecting the package. The second consideration lies in a smart combination of material and machine performance and strives to reduce the number of stoppages as much as possible, or in other words, to increase the "time between failures," which, depending on the type of machine, type of product and packaging material, and the complexity can be between 4 and 8 hours. It goes without saying that these values can be much lower (in the 30 to 60 minutes region) and that some in the filling and packaging industry have achieved times of above 8 hours, even 12 hours (Tetra Pak, Michael Grosse, personal communication, 2007).

In this context it should also be mentioned that really well-designed packaging, in combination with optimum materials, may help to shorten the normally annoyingly lengthy change-over periods, the periods when a new product is packaged on the same line with new packaging materials, which typically happens on confectionery packaging lines with flow wrapping.

The very last aspect of importance that we want to discuss in this chapter is the trend towards "hole-through-the-wall" solutions. This describes the situation in which a group of experts from a technology and/or ingredient provider works at the production site of the manufacturer. For instance, printer experts and machines are very often to be found in packaging environments; they are, as much as possible and meaningful, physically separated from the rest of the operations in order to ensure the confidentiality of other parts of the production. This has proven to be a rather successful undertaking and it does help the efficiency of the operations quite substantially. It is to be hoped that more of this will be done and converters, printers, and product manufacturers will work even more closely together in the future.

CONCLUSIONS

In Chapter 3 we discussed and elaborated on the following areas:

- Form and function in action; we looked at historic examples and discussed that deep understanding of the desired functionality of any object is crucial to successful execution.
- We showed examples from the early days of electronics design, such as personal computers and discussed this in the context of Apple and its head designer, Jonathan Ive.
- The importance of design in the consumer goods industry was discussed, with examples from the times when P&G introduced their approach to "open innovation" in the form of "Connect & Develop."
- The notion of "design thinking" (R. Martin) was briefly mentioned and introduced.
- We described the "perception of packaging" within the food industry and discussed the development of the perception of design and especially packaging design with the example of the Nestlé Company.
- The role of technology and design in the food industry, with special emphasis on food packaging innovation and renovation, was introduced.
- The "Seven Drivers of Packaging Innovation and Renovation" were discussed in great detail.

TOPICS FOR FURTHER DISCUSSIONS

- Which other historic examples, even ones that reach far back in time, can you find, and how do they relate to the topic of "form and function"?
- If this chapter's more contemporary design examples are too much skewed towards the Apple company, what other, numerous, examples can you find that support the idea of a combination of a deep understanding of any item and subsequent perfect and most appropriate design?
- Please find and discuss other examples of innovation and design in the consumer goods, especially the food, industry. Is design always given its true value in your company, organization, environment ?
- Please discuss your definition of "design thinking" and how this relates to your organization.
  Is the concept understood by your colleagues, peers, bosses?
- Based on your experience or understanding, what is missing in the concept of the "Seven Drivers of Packaging Innovation and Renovation"?
- How would your square the circle of (packaging) innovation at zero cost?
- How do you perceive the relative importance of 2D versus 3D packaging and how would you cope with the need or desire for increased legal and branding information on a package?
- For further detailed information on packaging drivers ("motivators") of relevance to the consumer goods industry, please consult:
  Market Motivators: The Special World of Packaging and Materials, E. Corner and F. A. Paine, CIM (The Chartered Institute of Marketing) Publishing, Chapter 8 (pp. 93–118) and Chapter 9 (pp. 119–141), Cookham, Berkshire, UK, 2002).

REFERENCES

H. Chesbrough (2003). *Open Innovation*, Harvard Business School Publishing Corporation, Cambridge, MA.

Cleantech-Switzerland (2012). http://www.cleantech-switzerland.com/en/index.php?page=1821C (accessed May 2014).

The Coca Cola Company (2014) http://www.cokerecycling.com/PET-Recycling-Video (accessed May 2014).

European Commission (2013). Eurostat Database, http://epp.eurostat.ec.europa.eu/statistics_explained /index.php/Packaging_waste_statistics and http://epp.eurostat.ec.europa.eu/portal/page/portal/waste /data/database (accessed May 2014).

EPA (2012). http://www.epa.gov/epawaste/nonhaz/municipal/index.htm (accessed May 2014).

C. Kotchka (2008). Innovation at P&G, Institute of Design Strategy Conference, May 2008.

R. Martin (2009). *The Design of Business/Why Design Thinking is the next Competitive Advantage*, Harvard Business Press, Cambridge, MA.

PET Resin Association (2012). http://www.petresin.org/recycling.asp (accessed May 2014).

4 Design and technology in academia: a new approach

…participation is key to the next big wave of innovation in business and society. Whether it is in the fundamentals of how we think about wealth or the economy, how we parse the minutiae of individual transactions, or how we evolve our most important social systems such as health care, I believe that the interconnectedness of our information society makes this shift inevitable and highly desirable.

Tim Brown, CEO and president of IDEO, and a thought leader on the subject of design thinking

FROM THE BEGINNING TO TODAY

Perhaps one of the most exciting areas of innovation today is found in academia. Never before has the world of education, and specifically design education, seen such radical disruption as in recent years as both technology and a generational mindset are redefining what learning and teaching is. Building on the knowledge from Chapter 2 "Design: from object to process," this chapter will focus on new approaches to design education that are challenging current models, redefining curriculum, and influencing the next generation of designers and innovators. Several factors that are driving new approaches to design and technology in academia include:

The vital role of partnerships—similar to the need for entrepreneurs finding complimentary partners, organizations creating multi-disciplined innovation teams and leveraging external partners for joint ventures, academic institutions, and programs must do the same, and can no longer work in silos.

Access to collaborative processes and tools—the availability and use of design-thinking methodologies, open-source platforms, and massive open online courses (MOOCs) are democratizing education and challenging academic institutions' current structures.

Understanding the value of business and academia—new approaches in education translates to new approaches in industry. Now more than ever both parties are recognizing the value they bring to each other—industry bringing real-world expertise and tools to

Food Industry Design, Technology and Innovation, First Edition.
Helmut Traitler, Birgit Coleman and Karen Hofmann.
© 2015 John Wiley & Sons, Inc. Published 2015 by John Wiley & Sons, Inc.

education, and academia delivering innovative processes and a creative workforce to organizations.

The rise of social innovation—the expanding role of design education, sustainability, systems thinking, and strategy is a catalyst for change and focusing on innovation for the greater social good.

The entrepreneurial mindset—education is moving from lecture-based learning to project-based learning. The hands-on learn by doing, fail-fast mentality is one of the tenets for design, as well as for entrepreneurs. Traditional learning, from primary education to higher education, is adopting this mindset.

Design process inspiring education at large—similar to the entrepreneurial mindset, design-based learning is beginning to be applied to all subjects at all levels. In the USA, the education system has fallen behind in comparison to other countries, with increased attention to STEM (sciences, technology, engineering, and math). The concept of STEAM (add "art/design" into the STEM) is one that will infuse design-based learning into our educational systems.

The Millenial generation's approach—discussion about new approaches in academia are not relevant without including some basic understanding of this demographic cohort. They have been driving and will continue to drive learning models and, in particular, higher education.

While new approaches to design education are developing at institutions around the globe, the perspective for this topic is coming from a designer and educator who is currently chairing the undergraduate product design department at Art Center College of Design. Examples of innovative and experimental programs and partnerships happening at Art Center will represent a modest percentage of exciting new approaches evolving all over the world. To understand the potential and power of where design education is going, it is important to contextualize where we have been, provide a brief history of Art Center College of Design and explain the unique relationship the college has with industry.

In the 1930s Edward A. "Tink" Adams, an ambitious advertising man with a vision of a new kind of education, opened and directed an art and design school in Los Angeles, California. His philosophy—teach real-world skills to artists and designers, and prepare them for business and leadership roles in advertising, industrial design and publishing. Instead of hiring academics to teach, Tink hired a faculty of working professionals from those creative fields to teach at Art Center, giving students education that is contemporary and relevant to industry and in turn giving graduates immediate employment options. The college was successful from the beginning and by the mid-1940s, once veterans were returning home after the war, developed a year-round curriculum and became an accredited four-year college. Art Center's renowned Transportation Design program played a significant role in attracting car manufacturers to open advanced design car studios in the 1950s, spurring the love affair with cars in Los Angeles. The college continued to grow and thrive, adding new disciplines in the 1960s and 1970s—photography and film, illustration, graphic design, and fine art—eventually moving the campus to a more spacious location in Pasadena, California, where the hillside campus still resides today.

Over the last several decades, the alumni from Art Center have had a tremendous impact on the design industry, shaping our culture with their vision. Studios inside corporations that have been led by these alumni include General Motors, Ford, BMW, Honda, Oakley, and Nokia, to name just a few. Ad campaigns like the "Got Milk" ad series, films like

"Transformers" and "300" and concept design for "Star Wars" and "Avatar" were created by Art Center alumni. Over the years, the academic leadership at Art Center has anticipated significant cultural influences, technological advances, and industry needs, responding with the educational tools and methodologies that remain at the forefront of art and design education. Art Center was the first design school to invest in computer labs, spearheading the revolution in digital design, and realized the importance of having a global view. From 1986 to 1996, the college had a second campus in Montreux/Vevey, Switzerland, developing European industry partners like Nestlé, Proctor & Gamble, and numerous European car manufacturers. Art Center continues to stay globally relevant today by sending students and faculty to work on projects across Europe, Asia, and Central and South America. In 2003, the college launched its groundbreaking *Designmatters* educational program focused on generating positive social change and improving people's lives through art and design, becoming the first design school to receive Non-Governmental Organization (NGO) status by both the United Nations Department of Public Information and the United Nations Populations Fund.

Today Art Center remains focused on the core educational mission of developing creative leaders and innovators in art and design, leading the way with cross-disciplinary programs, and studios that prepare students to make a positive impact in their chosen field, as well as the world at large. The current strategic plan created in 2012 leads with the statement "Learn to create. Influence change" representing the mission of Art Center College of Design moving forward. For more than 80 years, Art Center has earned an international reputation for its rigorous, cross-disciplinary curriculum, faculty of professionals, strong ties to industry, and a commitment to socially responsible design.

THE SPONSORED PROJECT: REDEFINING PRODUCTS, EXPERIENCES, BRANDS AND SYSTEMS

At Art Center, the curriculum is shaped around professional design practice and real-world challenges and methodologies. Courses are taught by working professional designers who bring their expertise and culture into the studio classrooms. Students learn by doing in project-based learning models, with courses typically lasting three to four months. Advanced courses are sponsored by corporations who are interested in the next generation's approach to a design challenge. What is delivered at the end of these projects reflects a thorough design process—from framing the problem or challenge, conducting research, identifying opportunities, exploring and visualizing numerous ideas, getting feedback from the sponsor, translating the research and feedback into well-formed concepts, refining and proving out those concepts, generating models that realize the idea, and delivering highly crafted presentations that communicate not only the form and function of an idea, but also why the idea is valuable.

The many sponsored projects that have taken place at Art Center over the years are as diverse as they are unique. They range from designing next generation consumer electronics to looking ten years out at the future of micro city cars, to peering 30 years into the future and designing pods for space travel. As described in Chapter 2, design as a process can be applied to just about any problem or challenge in the world—from redesigning simple products to redefining branded retail experiences or envisioning much-needed water delivery systems to impoverished neighborhoods. Companies like Honda, Nokia, and Nestlé have invested in numerous sponsored projects at Art Center, as the work has yielded fresh creative approaches to problems, delivered innovative solutions to real-world business challenges, and inspired the internal organizations involved to think differently about their businesses.

The Nestlé Corporation is one of Art Center's most valued partners, sponsoring a spectrum of projects over the past 15 years. The projects have included re-designing sleek Nespresso machines, packaging for the Buitoni pasta line, creating unique pet treats for Purina, developing premium chocolate products for Cailler, redefining the Wonka chocolate and candy brand, and envisioning future supply chains and retail experiences. These projects, along with sponsored projects with other brands, like Microsoft and Samsung, or manufactures, like Eastman plastics, make for great examples of design as a process of exploration and creating relevance. Art Center engages with both for-profit and non-profit partners, like NASA or the US Geological Survey, to leverage design as a process for creating meaningful solutions to some of the world's most difficult problems. More projects today are looking at the big picture, in need of systems thinking and collaborative methodologies to transform how business is being done or how design can actually transform the world. The following examples are organized in these three approaches that were introduced in Chapter 2 "Design: from object to process"—design as a process for exploration, design as a process for creating relevance, and design as a process for transformation.

Design as process for exploration

One of the most imaginative sponsored projects that took place at Art Center was with the Nestlé candy brand Wonka in 2010, called "*Feed Your Imagination: The Wonka Experience.*" The brand uses licensed materials for marketing and packaging the candy products, like Everlasting Gobstoppers and the Wonka Bar, taken from Roald Dahl's 1964 children's book "Charlie and the Chocolate Factory" and its two film adaptations, the 1973 film "Willy Wonka and the Chocolate Factory," starring Gene Wilder, and the 2005 film "Charlie and the Chocolate Factory," starring Johnny Depp and directed by Tim Burton. The challenge Art Center students were given by the leadership of the Wonka brand was: *To be the most-loved and inspiring confectionary experience in the world.* The goal for the project was to re-imagine the future of the Wonka Experience based on the stories from Dahl's book and the two films. Wonka wanted to move from the current line of sugar brands (Sweet-Tarts, Nerds, Laffy Taffy) and Wonka chocolate bars, targeting tweens and young adults, to a mega-brand targeting all of those that are young at heart and desire the Wonka Experience. Students were to create concepts that elicit "only Wonka could do something like that!" from their target consumers, meaning that anything created had to have an obvious visceral and emotional connection to the whimsical Wonka story. This was a design students' dream project as they were asked to define the entire Wonka Experience—from the product and packaging to the merchandising and point of purchase to the retail environment for Wonka confections. A strong emphasis was placed on storytelling, ritual, discovery, and the consumer experience, driven by the Wonka tag line "Feed your imagination." Teams of Art Center students from product design, graphic design, and environmental design were required to read the book, watch both films and provide critical comparisons (essays and visual analysis) of the story, the eccentric characters, and the sensual environment of Wonka's chocolate factory, including the floors, walls, ceilings, lighting, textures, surfaces, materials, colors, sounds, smells, and objects. Other research exercises included:

- Exploring and benchmarking confectionary products and packaging and analyzing their sensorial qualities, including appearances, tastes, textures, and sounds.
- Observing a chocolatier at Nestlé making chocolate confections, and understanding the science and the process of how chocolates are made.
- Creating descriptive words based on the attributes of the five child characters in the book (devilish, charming, arrogant, nauseating, malicious, loud, humble, etc.), and inventing treat words associated with those characters, like everlasting gobstopper, crumpscoddle, wraprascal, humplecrimp, snozzcumber, etc., followed by imagining and visualizing in sketches or models what the ingredients of these treats might be.

What resulted after 14 weeks of exploration and concept development was a series of in-depth proposals created by Art Center trans-disciplinary teams. Ideas and solutions that "fed the imagination" included the "Diptastic Creamery," an in-store chocolate bar/lounge, the "Wonkatronic Confectionizer," a vending 3D printing machine making personalized chocolate confections, and a collection of gastro-chemical inspired confections like "Fizzlefazzles" and "Glumptious Gigantecules". One of the teams, "Loompa Labs," based on of Willy Wonka's factory workers, the Oompa Loompas, went as far as designing and wearing the white lab-coat-like uniforms to be worn when entering the clean factory retail experience. The students visualized a series of whimsical machines so that in-store consumers can watch their "Loompaliscious" confections being made in front of them. The Loompa Lab team actually created a dozen different kinds of uniquely shaped and flavored confections, inspired by their research experiments making candies with a professional chocolatier.

The success and sometimes overwhelming creative results from projects like the Wonka Experience are due to the commitment of the sponsor. There were several representatives from Nestlé that were very involved in the entire project, providing the students with consumer insights and trend information, and access to confection making, packaging, and brand expertise, as well as food science knowledge. The students integrated all of that inspiring knowledge into their design process, and at the end presented ideas to Nestlé leadership that they could have never imagined. The magic and beauty of having a group of young designers working on a project like this is that they have an unbiased perspective of the brand and are not bound by the current technologies, manufacturing constraints, or marketing strategies that drive the current states of corporations. They are free from the day-to-day internal challenges that companies have, to innovate, explore new ideas or material applications, and re-imagine entire brand experiences. For design education, this collaboration is critical as it provides students with access to knowledge outside of design and academia, but rather from the real-world business and technology sectors. For the sponsor, this provides a source of inspiration for their innovation teams that no focus group or marketing survey could possibly give, and gets corporate leadership to think differently about the future of their business.

While the full semester-long sponsored project has proven to be successful over many years, there are recent new collaborative models that have been developing at Art Center. A process that a group of faculty has developed and trademarked that embodies a design-thinking approach is the DesignStorm™. The DesignStorm™ is a one- to three-day immersive innovation session that involves Art Center students and faculty from different design disciplines, along with representatives from the sponsoring organization in tackling a specific challenge or exploring new market opportunities. The challenge can be driven by a new material technology in need of innovative applications, or an opportunity to re-think product experiences from manufacturing, to shelf, to user experience. Regardless of the challenge specifics, the iterative process is constant—learn by doing and making.

This model and process has opened more doors for new partners from the raw materials manufacturing world to collaborate with the Art Center community, like Avery Dennison, PolyOne, and Eastman Innovation Lab, specifically exploring innovative materials applications to re-imagine manufacturing methods, consumer products, and packaging. A great example is a DesignStorm™ with Eastman Innovation Lab where the over-arching goal of the project was to re-think how hand-held consumer electronic devices can be created utilizing Eastman's Tritan co-polyester material. The properties of Tritan can make it a viable drop-in replacement for polycarbonate in many cases, presenting designers with an opportunity to innovate with fewer constraints. The objective of the DesignStorm™ was to re-think what cell phones, digital cameras, games, game controls, and other hand-held electronic products could look and feel like, if made with the Tritan material. A dozen Art Center students along with material scientists and experts from Eastman spent a day together, quickly understanding the co-polyester properties and then letting the students' imaginations loose to explore new applications. Through the design-thinking, rapid visualization, and prototyping process, dozens of unique ideas were presented at the end of the day, including rough 3D prototypes that the sponsor could touch, feel, and hold. In the end, a student was given the opportunity to continue work on the project at their corporate headquarters as an intern—another benefit for both parties of the sponsored project model.

Just as this model has opened doors for material manufacturers to design thinking, technology companies like Microsoft are also finding the benefits of collaborating with Art Center in the DesignStorm™ process. During summer 2013 the Surface tablet team from Microsoft collaborated with a group of Art Center students from product design, graphic design, and interaction design in a three-day DesignStorm™, with the goal of coming up with new and unique concepts for the "blades" accessories that can connect to Microsoft Surface tablets. The Surface team leadership wanted to capture the process to show as part of their press event for the premiere of the Surface 2 and Surface Pro 2 tablets, the intention being to inspire people to want to make their own versions of the Blade, and to show the designers in the process of brainstorming and making their ideas come to life. After three intense days in the studio, fueled by music playing and lots of food, the final results were impressive, and within one month the Microsoft team showed a video capturing the DesignStorm™ at the Surface 2 launch event in New York City. Concepts included an array of funky new game pads and artists' pads, music keyboards, educational products, and a solar-powered panel for the tablet. DesignStorms™ like this are an effective way of companies getting a glimpse into the future of how the next generation of technology consumers will be defining their product experiences.

Design as a process for creating relevance

As described in Chapter 2, design is no longer just simply and arbitrarily styling products, but defining the entire experience and creating an emotional connection between people and their products. The example given of creating relevance in that chapter was the Samsung-sponsored project at Art Center challenging students to explore the expression of human emotion in colors, materials, and forms in future mobile communication experiences. The relationship between humans and their products begins with "visual first impressions" that is often dictated by the surface color and materials. The project revealed how emotion can be expressed in a product's surface and how ubiquitous products like a phone can be differentiated by their skin, creating meaning for specific users. Results from the project demonstrated how important it is to understand deeply what drives specific consumers, beyond assumed needs and functional features, and to translate their desires into products that have emotional relevance.

In 2009, Nestlé came to Art Center asking a similar question—how to define "super premium" Nestlé-branded experiences that will set them apart from their competitors. The challenge given was to first define what exactly "super premium" was, at a time when the idea of luxury and luxury products was being re-defined after the financial crisis in 2008. The language and approach was shifting from "luxury" to "premium." In order to understand how to create premium products, it was critical to first define "premium" and then apply the language to a brand and a line of products. This approach led to a two-semester engagement between Nestlé and Art Center: a three-day "Super Premium" DesignStorm™ and the following semester a full-term project working with the Nestlé Confections team on "The Supreme Chocolate Experience," targeting Nestlé's Cailler chocolate brand.

The objective of the DesignStorm™ was to take a broad approach to the topic by first identifying needs and discovering opportunities to re-define premium packaging and experiences for the food and beverage industry from a North American perspective, with the deliverables being visual narrative strategies that Nestlé would exhibit in their headquarters in Vevey, Switzerland. The idea was that the "super premium" language developed by the students could inspire and influence multiple business units at Nestlé. The DesignStorm™ began with the Nestlé team and Art Center students sharing their perspectives on trends, insights, and their current definitions of "super premium." Out of the knowledge sharing and discussion came a series of opportunities identified for the student teams to ideate and develop a new language—both in words and visually—with rough concepts demonstrating the future of super premium products and packaging. Categories explored included beverages, desserts, ice cream, coffee, frozen foods, pet food, and of course, confections and chocolate. The students defined the language of super premium with a focus on authenticity, sensory and emotion, personalization, and exclusivity. An exhibit and a presentation were developed after the DesignStorm™, and displayed and shared with the innovation teams at Nestlé headquarters. The results from the DesignStorm™ also became the foundation for "The Supreme Chocolate Experience" project the following term.

Along with the language from the "super premium" DesignStorm™, students were given the history of Nestlé's premium chocolate brand Cailler—founded in Gruyere, Switzerland by Alexandre-Louis Cailler, and the oldest Swiss chocolate brand. The challenge given to Art Center students was to create a "supreme chocolate experience" that will reposition the Nestlé brand as the number one chocolate and confectionery maker in the world. Emphasis was placed on creating innovative packaging and retail experiences that reflect the ultimate premium branded experience. Like the DesignStorm™, the results from this project were also displayed at Nestlé in Switzerland, but this project was exhibited at the opening of Nestlé's Chocolate Centre of Excellence (CCE) in Broc, Switzerland. The CCE highlights Nestlé's chocolate-making heritage and drives innovation in the premium and luxury chocolate segment. The centre is home to a team of Nestlé chocolate specialists, cocoa-bean scientists, sensory experts, chocolatiers, packaging designers, consumer specialists, and a panel of independent chocolatiers, providing artistic inspiration. This spirit of open innovation is furthered by the Centre's partnership with prestigious external design institutions such as Art Center College of Design and the ECAL University of Art and Design in Lausanne, Switzerland. The work coming out of "The Supreme Chocolate Experience" included new forms and shapes of chocolate products, numerous innovative packaging solutions, and compelling graphics and point-of-purchase displays, as well as fresh marketing strategies reflecting the language of "super premium." Several of the students from both projects went on to internships at Nestlé, continuing development of the work, with a couple of them getting hired by Nestlé upon their graduations. The fact that both Art Center projects were exhibited internally at Nestlé and the students were eventually hired as designers at Nestlé is a testament to the successful and meaningful partnership between global corporation and educational institution.

Both the Samsung and Nestlé projects are very good examples of design creating relevance in the making of consumer products for mass consumption. But what about design creating meaningful solutions for those people that need innovation the most? In developing regions of the world that are challenged in having access to the basic needs of every human—clean water, nutritious food, protective housing, and education—design is beginning to play a bigger role. This is where design meets social innovation and where designers today have the biggest opportunity to impact the lives of people all over the globe. *Designmatters* at Art Center was mentioned in Chapter 2 and the opening of this chapter in the overview of Art Center's history and impact. It is an initiative that began a decade ago, driven by a small group of Art Center leadership, staff, and faculty that saw the need for design and design education to get involved with local and global projects that focused on design creating positive change. In its mission, founder and vice-president of Designmatters at Art Center, Mariana Amatullo, states "Designmatters allows us to look at the world as a classroom with an eye toward changing it for the better. We aspire to redefine and expand the role of the artist and designer into one who is a catalyst for social change and innovation."

The vision and mission of Designmatters (DM) is to provide the know-how and aspiration to shape the futures we truly desire for a more sustainable and equitable world, by inspiring, educating, and forming the next generation of creative leaders. Through research, advocacy, and action, DM engages, empowers, and leads an ongoing exploration of art and design as a positive force in society. DM is integrated across all the educational departments at Art Center and active at three key levels: as an educational magnet and research division for the college, DM conceives of projects for the curriculum, oversees the DM Concentration at the undergraduate level, and partners with Graduate Media Design Practices in the MDP/Field Track; as an agent for social impact educational projects, DM is a guarantor for implementation and assessment of projects led by students, faculty, and alumni; as an external relations center for strategic partnership building, DM leverages art and design education as a tool for positive change in the world. Over the last decade, Designmatters at Art Center has engaged in numerous projects, tackling the most challenging issues we face as a society, from poverty and homelessness, to gun violence and abuse, to health and education. One example that best illustrates the power of design to make change is "Safe Agua"—a series of projects collaborating with non-profits in South America to design and implement systems for providing access to clean water in impoverished communities that currently do not have running water.

Safe Agua projects have taken place over the last five years in the "asentamientos" and "campamentos" (slums) of Santiago (Chile), Lima (Peru), and Bogota (Columbia), in partnership with their local non-profits such as TECHO, a prominent Latin American NGO that serves communities living in slums and Socialab, the Innovation Center affiliated with

TECHO. Leaders from these organizations, along with the DM leadership and Art Center faculty, identified specific locations with a cross-section of families who live in conditions of extreme poverty and are affected by systemic challenges due to lack of access to running water services. Safe Agua projects aim to gather human-centered data and develop design prototypes for services and products co-created with a community of families living in extreme poverty. Each of the projects began with an immersive two-week field research trip to the South American communities, where students worked alongside families and community partners to uncover needs and identify design opportunities around water and to help overcome the cycle of poverty. The goal of field research is to establish empathy and deep connections with the local families, and gain understanding of people's dreams, needs, and constraints. This can only be done through participation and direct experience with the families, asking qualitative and quantitative questions, and listening to people's stories. Driven by the field research, the students analyze the data to inform their conceptual/design development phase, with the goal to create complementary, low-cost prototypes and systems to address the specific water-related needs they observed with families on-site. After the two-week field research trips, students returned to Art Center to continue developing the projects. The co-created design concepts, first ideated during the field trip, are tested, developed, reiterated, and refined. This process of development is supported by lectures from local social innovation experts, business developers, entrepreneurs, and the projects' non-profit partners. Halfway through the projects, members of the team returned to the South American communities to test the prototypes in progress with families. Using the feedback gained from the field tests, the students incubate and develop the projects further for final presentations. Given the user-centered approach, the Safe Agua projects aspiration is for teams to develop design solutions that can improve the lives of those living in extreme poverty, and that through the private/public sector alliances, opportunities may emerge to further develop the outcomes into market-based opportunities for the Base of the Pyramid.

The Safe Agua projects have produced a portfolio of innovative ideas, concepts, and product solutions that have provided the asentamientos and campamentos access to clean water, potentially improving the quality of life in these communities. Examples of these innovative solutions, extracted from the project descriptions on Art Center's Designmatters Safe Agua websites, include:

Ducha Halo is a shower product delivering hot water bathing to people living with no running water and inconsistent electricity. Ducha Halo's 6-liter metal container (with an analog temperature sensor) is easy to heat on the stove, reducing preparation time and physical strain. Two minutes of hand pumping yields a 15-minute hot shower, which flows in a refreshing, generous stream from a cheerful plastic showerhead.

Balde a Balde ("Bucket to Bucket") is a portable faucet that provides running water from any container, bringing the health benefits and experience of using a tap to families living without running water. The user attaches Balde a Balde to any container with a universal clip, then begins a continuous flow of water with just a few squeezes of the siphon pump. Users can easily control the exact amount of water they need, with a simple click of the on/off spout or a twist of the valve to regulate flow. Balde a Balde harnesses gravity to bring the dignity of running water to the three billion people living without taps.

WaterWays: Balde Movil. Without running water, hand-carrying water in and between homes is physically straining, inefficient, and makes water prone to contamination. BaldeMovil in an inexpensive, lightweight dolly that makes it easy to transport two common 5-gallon buckets, along with other heavy items, even when rolling on hilly terrain. This innovative product makes water accessibility within the domestic sphere more efficient, more sanitary, and less physically stressful.

GiraDora is a human-powered washing/spin-dryer a user can sit on that increases the efficiency and improves the experience of hand-washing clothes. For under $40, GiraDora more than doubles productivity, increases the health of women and children, and affords the opportunity to begin breaking the poverty cycle. The user sits on top of the drum-like appliance and pumps a pedal with her foot, which agitates, cleans, rinses, then spins-dries clothes. While providing a more comfortable, ergonomic, and efficient way to clean clothes, GiraDora also affords opportunities to generate income.

These Safe Agua projects have yielded award-winning design solutions and have been acknowledged by organizations around the world for the human-centered innovative collaborative process. Recognition has been received from the International Design Society of America, Industrial Design Excellence Awards, Core77 Design Awards, Spark Design and Architecture Awards, the National Collegiate Innovators and Inventors Alliance, and publications like *Fast Company* and *Business Week*. Beyond the accolades, interdisciplinary student teams are continuing to work towards sustainable, scalable, real-world implementation of these projects, which are in various stages of development. Design students, along with business partners from Society and Business Lab at USC Marshall School of Business, are continuing the development of sustainable business plans, including sourcing, pricing, marketing, and identifying potential partners/manufacturers in order to bring the products to market.

Design as a process of transformation

The final example of Art Center projects demonstrates the power of partnerships, collaborations, and shared vision for design as a process of transformation in creating a better world. Design in this form is not intended to create products or refined imagery, but rather is a way to convene large groups of people with very diverse views to think strategically about complex issues, define scenarios, and visualize new futures. The example described below was no simple feat and required a dedicated team of people, Art Center leadership, staff, faculty, and expert partners, and weeks of organization and planning to orchestrate these large-scale design-thinking-focused gatherings.

Art Center's campus in Pasadena, California, is in an ideal location to convene the world's top scientists and technologists in the fields of earth and space sciences, with expertise in weather and ocean patterns, atmospheric conditions, and climate change. Art Center is neighbor to the California Institute of Technology (Caltech) and the Jet Propulsion Laboratory (JPL), and over the years, Art Center has collaborated with these innovative institutions on a variety of projects. In 2010, following the world's latest natural disaster, the earthquake and tsunami hitting Haiti, Art Center Designmatters partnered with the Multi-Hazards Demonstration Project (MHDP) of the United States Geological Survey (USGS) to develop a series of events focused on awareness and response to these devastating storms. In recent years the USA has dealt with Hurricane Katrina and Superstorm Sandy, both killing hundreds, wiping out cities and neighborhoods, and costing millions of dollars in recovery. The experts at USGS predict that "superstorms" and epic earthquakes are on the rise and that what happened in Haiti, or in Japan in 2011, could happen in California. The disaster in Haiti killed over 200,000 and four years later there are still residents living without electricity and running water. The USA has not yet experienced a natural disaster on that scale, but the USGS predicts that we will. How can that kind of knowledge and data inform public awareness of the magnitude of superstorms, encourage the people to be prepared, and better synchronize emergency response agencies when a disaster like this does hit?

Over the span of six months a trans-disciplinary team of Art Center faculty, students, and scientists from the MHDP developed a series of interconnected events and outcomes that would bring the superstorm scenario to life. The ARkStorm is a hypothetica, but scientifically realistic superstorm scenario that was developed by the USGS (prior to this partnership) describing an extreme storm, the Atmospheric River 1000 Storm, named after a way of quantifying the magnitude of storms and a parallel to the biblical Noah's Ark story. This kind of epic storm could devastate areas of California, causing hundreds of billions of dollars in damage and recovery, and displacing a quarter of the state's population. With no pun intended, the "DesignStorm™" process was an ideal method used to gather and inform numerous stakeholders and participating organizations around the ARkStorm scenario.

The first ARkStorm DesignStorm™ was a one-day event held at Art Center and included a broad range of participants—emergency managers, structural engineers, meteorologists, business leaders, geologists, social scientists, and policymakers. Participants worked in short, immersive sessions throughout the day, in teams with Art Center faculty and student "visualizers," and in pre-assigned groups designed to maximize expertise on the following key issues around ARkStorm: communication strategies and community involvement, emergency response planning and evacuation procedures, infrastructure planning and flood control, storm scaling and consistent messaging. The goal was to craft future scenarios and possible solutions that could be driven by the stakeholders in the room, and at the end of the day over 1000 ideas were generated with 100 plus critical topics identified. Vital connections were discovered through this process, giving momentum to developing a larger strategic engagement—the ARkStorm Summit.

The US Geological Survey (USGS), Federal Emergency Management Agency (FEMA), and California Emergency Management Agency (CalEMA) hosted the ARkStorm Summit in January, 2011, a two-day event to engage stakeholders from across California to take action as a result of the USGS ARkStorm Scenario's findings, which were officially released at the Summit. The Summit, facilitated by Art Center faculty and students, drew over 200 participants and leaders from across the state and the country, from 65 agencies and organizations. Emergency managers, regulatory and scientific agencies, policymakers, business leaders, and other experts from the public and private sector, were engaged to address data surrounding the real potential for a massive storm, and to convene on actionable interagency steps moving forward. To encourage dialog, all participants participated in interactive breakout sessions led by Art Center facilitation teams, leveraging similar design-thinking methods to those used in the first DesignStorm™. What resulted from both of these events was a communication strategy, a public awareness film, and a process for these vital cross-sector decision-makers to conduct emergency planning meetings. Since the ARkStorm projects, the agencies involved have continued to leverage design-thinking tools to collaborate and improve their systems of communication and emergency response preparation in the future.

THE EXPANDING ROLE OF DESIGN/BUSINESS BEING DESIGN DRIVEN + DESIGN BEING BUSINESS DRIVEN

One of the most critical partnerships for design innovation is with business, and as the international marketplace has embraced and valued design over the recent years, there have been more opportunities for designers to play influential roles in business development. With the emergence of design thinking in MBA programs and the corporate world, it is now common to see corporations integrating designers into their leadership and design organizations integrating business executives into their teams. Such opportunities require a deeper understanding of the intersection of design and business, as well as the ability to understand each other's languages and collaborate across disciplines. A decade ago, leadership from Art Center College of Design and INSEAD international business school understood this need and launched a visionary pilot program to build awareness within tomorrow's industry leaders of design's strategic role in creating sustainable, top-line growth. In turn, the value to Art Center's design students was their active participation in INSEAD's MBA curriculum and the practical insights they gained into the competitive issues that impact business. This collaborative initiative has been well received by students, professors and administrators from both campuses over the last ten years, with the intention of continually improving the program, and runs once a year at INSEAD.

INSEAD, known as "The Business School for the World," is widely recognized among top-tier business schools across the globe as one of the most innovative and influential. It is the only business school that spans three continents with full-fledged campuses in Asia (Singapore), Europe (France), and Abu Dhabi. INSEAD's unique global perspective and multicultural diversity are reflected in all aspects of its research and teaching. Their business school partnerships include the Wharton School of the University of Pennsylvania, the Kellogg School of Management at Northwestern University, and Johns Hopkins University. Art Center College of Design and INSEAD, two of the world's leading educational institutions, have taken their first steps toward bridging the gap between the quantitative drivers of business and the qualitative character of design. Once a year, ten students from Art Center are selected to attend INSEAD (the program alternates between Singapore and France) for a semester, to collaborate with MBA students on a product development project and take a series of MBA courses. Halfway through the semester, during a one-week break, MBA students fly to Pasadena, California to get immersed in the world of design, with workshops at Art Center and tours of local design and innovation studios. The goals of the program include promoting the collaboration between design and business, preparing students for the global economy, educating students to become design leaders, and understanding new areas of design application. This special program has shaped design students into strategic-thinking leaders and MBA students into creative-thinking leaders. Since the inception of the program, many of the alumni from both institutions have fast-tracked careers in design strategy or innovation organizations, leading internal corporate innovation groups or consultancies. Many have launched their own ventures and recently graduates from INSEAD and Art Center that collaborated in school are now collaborating in industry, some launching new companies together.

The partnership with INSEAD gives Art Center students a unique education that they could not get in Pasadena and INSEAD students a taste of the design innovation process. It is a powerful academic partnership, benefiting both institutions and their graduates. Later in this chapter there are more examples of academic programs focusing on the intersection of business and design. Another kind of partnership Art Center is developing with business schools is the creation of new executive education programs. As design thinking gains traction as a valued innovation process and business tool, the opportunities for Art Center's collaborations with business schools continue to grow. Several years ago, prior to the financial crisis, Art Center developed an executive education program with ESADE business school in Barcelona, Spain, a global leader in executive education. A small team of faculty at Art Center and ESADE developed "Beyond Pretty," a series of hands-on creative sessions that ran over the course of a week, inspiring business leaders that design was more than just making things look pretty, but rather was a process for innovation. The purpose of the series was to teach corporate leaders, senior executives, innovation managers, and entrepreneurs the power of design thinking, and transform today's business problems into opportunities for radical disruption. Tomorrow's business leaders must embrace new kinds of risk-taking in order to seize distinctive market propositions that enable their companies not only to survive, but to thrive. The "Beyond Pretty" program taught the skills executives needed to bring new creative solutions into their companies, using design-thinking processes as tools for corporate innovation and transformation.

A team of Art Center faculty and students facilitated these innovative programs in Barcelona with success, even during the years of the financial crisis, sharing creative design process techniques and methods through a project-based learning model. The design students transformed into "coaches" and "visualizers" for the teams of executive participants (truly a unique real-world learning experience for students) as Art Center faculty led the group through an overarching learn-by-doing innovation challenge. Through a series of creative exercises the executives learned: how to visualize information, design research, and critical-thinking methods, ideation and prototyping, how to create compelling future scenarios and develop innovative new venture propositions.

FROM "MULTI-DISCIPLINE" TO "ÜBER-MULTI-DISCIPLINE" AND THE FUTURE OF "DESIGN +"

The examples of collaborative projects and partnerships describing design as a process for exploration, creating relevance, and as a process for transformation demonstrate the shift in design as an industry and in design education. Design as a discipline by itself is ineffective and the same can be said for technology and business in today's highly competitive world. Once a differentiator at Art Center, the trans-disciplined or multi-disciplined sponsored project has been adopted by design programs all over the world and is now becoming expected by the industry. Depending on the design college or design program within a university, what being "multi-disciplined" means can vary. For large universities, it means that schools of design can collaborate with schools in engineering and business. For colleges like Art Center, the disciplines are just art and design related. These small colleges rely on external partnerships to provide expertise in the fields needed outside of design. Our education now must move from being internally multi-disciplined to being "über-multi-disciplined". While that is not an easy thing to organize and execute, ideal education and projects yielding true innovation will require the collaboration of design, business, social science, technology, and engineering.

Several years ago Tim Brown, the CEO and president of IDEO and one of the pioneers of design thinking, articulated the idea of a "T-shaped" designer. The shape of the "T" represents the depth and breadth that designers need to possess in order to successfully innovate in the world today. The vertical stroke of the "T" represents the depth of creative and making skills and the horizontal stroke of the "T" represents the ability to have empathy for and collaborate across disciplines. While there is some controversy in the design education community that design risks losing its credibility if it gets stretched too thin, having empathy is critical, because it allows designers to imagine problems from another perspective and fosters collaboration. The idea is that designers can be effective facilitators of innovation if they have deep expertise in their field and are familiar, somewhat knowledgeable and appreciative of other disciplines.

If we believe that future creative leaders are indeed "T-shaped" and are able to navigate, participate in, and facilitate "über-multi-disciplined" innovation teams, then design education needs to evolve. This is not an easy task, as the biggest challenge for design education today is balance and compression. It is imperative that design schools ensure students are getting the very best education in order to prepare them for the real world, yet the number of critical skills needed for success, and in demand by the industry, continue to rise. Design education is now very "additive," with very little "subtractive" in its curriculum. The ability to develop curriculum that maintains the depth of creative and making skills designers have to have in order to earn their titles, but also ensuring that students have a breadth of horizontal skills coming from other disciplines (including well-developed interpersonal communication and collaboration skills) is extremely challenging to build into a four-year undergraduate program or even a two-year graduate program. These challenges are forcing us to restructure design education, integrate new models of learning, and encourage the value of self-reliance and life-long learning to our students.

Because of these issues of depth and breadth that designers are facing, perhaps design needs to think about encouraging and preparing students to integrate a "+" into their educational careers and professional development. Many of Art Center's most successful alumni, as well as alumni from other design schools, have either had prior degrees in something else outside of design before going to design school or end up getting an advanced degree in business, social sciences, or a technical discipline like engineering or computer science after design school. The design degree provides creative and technical depth, the "+" influences the approach to design, adds breadth to a designer's perspective, and obviously contributes to a designers ability to "specialize." This "design +" approach to education can be attained in many ways—either through traditional university degrees, continuing education, technical training, executive education, or self-learning via online classes. Regardless, it is important that designers and students have a long-term view on their careers, mixing "generalist" skills with "specialist" skills. Design alone may not be enough anymore to be competitive in today's world. The tenants of "Design +" include the following:

Integrating the "COs": collaborate/co-create/co-locate/co-curricular

From the numerous project examples above, along with design needing to be "T-shaped," it is apparent that design does not work in isolation. It is critical that students and young designers understand that design is not about working in a bubble. Education could benefit and have far greater impact if academia could focus beyond the classroom and integrate the four "COs" into the educational experience:

Collaborate—integrate team-based projects into curriculum with partners from industry

Co-create—design *with* not just *for by* involving consumers and end-users into the process

Co-locate—get students out of the comfort zone by expanding cultural horizons via study-away programs

Co-curricular—adding educational value beyond the classroom and into culture.

Design + technology + business = the DNA of innovation

These three ingredients are critical for innovation and for educating future innovators. Each of the disciplines can vary in scale and emphasis on overall curriculum and in projects, but design, technology, and business all have to be considered in teaching and practicing innovation. Innovative products, services, and experiences are derived from a combination of:

Desirability—human-centered approach, user empathy, and aesthetic details (design)

Feasibility—functionality, manufacturability, and successful execution (technology)

Viability—marketable, profitable, and sustainable (business).

Innovative spaces for learning

Just as critical as the cross-disciplined curriculum are the environments in which the learning is taking place. With participatory learning the traditional lecture-hall classroom is dead. Spaces have to be flexible, compelling, and feel more like an experimental lab or a cool lounge to inspire innovation. Art Center has been investing in this idea over the last few years and has a benchmark that reflects this approach. The Color, Materials and Trends Exploration Laboratory (CMTEL) was founded in 2006, with the generous support of $2 million from the Nokia Corporation, in order to provide artists and designers with the resources to become better problem-solvers, innovators, storytellers, and leaders. Art Center has developed an innovative program and laboratory space to provide students with expertise in color and materials technologies, as well as understanding global trends. Designers today are asked to come to the job market with a broader range of skills than ever before. In addition to the technical and creative skills expected, graduates need solid analytical and research skills to excel in their professions. The CMTEL is a ground-breaking facility that provides Art Center students with all of that and more, placing Art Center and its student body at the forefront of color trends, emerging material technologies, and a variety of market trend information.

Since its opening, CMTEL has hosted events, lectures, and courses that focus on color, materials, and trends. The lab brings professional color and materials designers to share their experience and work methods with students, offers courses on topics as varied as sustainability, lighting technology, and textile manufacturing, and sponsors projects in which undergraduate and graduate students are given the guidance and the facilities to explore color, materials, and trends at the deepest level. CMTEL is the founder and home to Art Center's DesignStorm™ process, hosting numerous immersive, innovative sessions with corporate partners every year. CMTEL's mission is to create a venue for students to visualize the future, giving them the tools and processes they need to turn their projects into reality. The lab serves as an integral part of the College's distinctive design sciences and research curriculum, providing future designers with a combination of technical, analytical, and research skills, as well as a comprehensive understanding of materials technologies, color application, global trends, sustainability, design research/strategy, and business processes. CMTEL is committed to making all of Art Center's students—from illustrators to industrial designers—the best problem-solvers, innovators, and decision-makers in their fields.

Beyond the programmed educational benefits CMTEL delivers, there are other benefits that are by-products of innovative spaces. Without prompting from faculty at Art Center, the students have used the CMTEL as a social gathering spot. This has encouraged spontaneous collaboration, knowledge sharing, and peer-to-peer mentorship. The lab also became a "clubhouse" as it hosts weekly meetings for student clubs like the "EcoCouncil" or the "Concept Help" work sessions. CMTEL has contributed to Art Center's creative and intellectual culture in ways not imagined when it was conceived. This generation of students craves collaborative spaces where they can do research, work on projects together, and simply share ideas over a cup of coffee.

PREPARING THE NEXT GENERATION OF INNOVATORS/ THE "EXPERIENCE PORTFOLIO"

As we plan for the future of educating "Design +" students, it is important to understand the mindset of the Millennial Generation that is currently in higher education, will continue to enter college and enter the workforce for the next two decades. Numerous books and articles have been written on the subject of the Millennials' attitudes, behaviors, and motivations, and their eventual impact in the workplace. According to the demographic and generational change experts, Neil Howe and William Strauss, authors of *Millennials Rising: The Next Great Generation*, this cohort includes people born between 1982 and 2002, currently spanning teenagers to those entering their 30s, soon to become the dominant cohort in the American workforce. From the perspective of leading a program in higher education, there follows a brief summary of the DNA of the twenty-something student and observations on this Millennial Generation:

- They were digital at birth, they are the first generation in college that do not know life without a computer or even a smartphone.
- They are gamers, not tinkerers, as they spent time with an XBOX, Nintendo, or Wii, not with dad fixing cars or mom in the sewing room.
- They are typers not writers, and the act of putting pencil to paper is more foreign than thumbing away texting on a smartphone.
- They are "hyper-taskers," not just multi-taskers, frequently playing online games, texting, listening to music, and web chatting, all at the same time.
- They are thought to be entitled and feel they should be rewarded just for participating.
- They hold high expectations for their academic and work experiences.
- They are optimistic, with the majority wanting to make a difference in the world.
- They are the "open-source generation" as they thrive and excel in collaborative work cultures versus competitive environments.
- They have an entrepreneurial spirit, most wanting to be their own boss and have flexible work schedules, and are naturally integrating work life with personal life.

Without getting into the specific challenges and benefits that the Millennials bring to education and the future workforce, design education has an opportunity to better prepare the next generation of innovators. As mentioned previously, the challenge of compression and balance, primarily in "hard skills" (technical, creative, making) is coupled with the need for design students to build their "soft skills" (social and professional). It is critical that students graduate from design school with more than just a traditional portfolio, but with real experience that employers can be confident in hiring them to deliver. The "experience portfolio"—team projects, presentation skills, leadership, internships, and professionalism—is just as important as the traditional portfolio showcasing projects, and creative and technical skills. Now more than ever, employers are hiring talent based not just on their design skills, but on their character and emotional intelligence. Companies are looking for designers who know how to collaborate, be a team player, be a leader, be accountable, and can easily integrate into a professional environment. Design managers simply do not have the time and resources to educate new hires on the nuances of professional life. By integrating co-curricular learning experiences into design education, through internships, leading a student club, or finding mentors, students will gain those soft skills needed to succeed into today's competitive world.

NEW VENTURES IN DESIGN EDUCATION/FROM NON-PROFIT TO FOR-PROFIT

Over the last few years, as described in Chapter 2, we are in the midst of the "Kickstarter" culture, with numerous crowdsourcing platforms at the fingertips of every designer. It is the responsibility of design schools now to integrate entrepreneurship and startup business knowledge into their programs. This generation of design students is engaging with these platforms regardless of how much they think they know. It is critical for their success that they are armed with the proper knowledge (business, technical, and legal) and mentors that can advise them on what to do and what not to do. Taking a page from the tech world and tech-driven academic programs, Art Center is launching an entrepreneurial studies program and participating in the start-up culture of accelerators and incubators popping up all over the globe. In recent years, as Art Center's academic connections with Caltech have grown, students and faculty from both institutions are collaborating in social and business entrepreneurship courses and projects. Caltech students are in search of good designers to partner with, and design students realize they need the technologists to bring their ideas into reality. The Design Accelerator (TDA) is a partnership between Art Center and Caltech and in-residence at Idealab, a business incubator in Pasadena, California founded by entrepreneur Bill Gross; TDA promotes cutting-edge design, technology and business strategy in creating start-ups, reinforcing the "innovation trifecta": design + business + tech.

The Design Accelerator is a for-profit entity, outside of the non-profit Art Center academic institution, focusing on early stage design and technology businesses, with opportunities to use great design as a critical disruption point. The Accelerator bridges the gap between students and alumni having great ideas just sitting in their portfolio, or rely solely on Kickstarter, by giving them an opportunity to develop their concepts into real commercial ventures. Businesses are selected for TDA based on their ability to demonstrate the value of human-centered design in support of advanced technologies and the potential to disrupt the marketplace. Start-ups in the first class include a foldable electric bike company, a healthcare company focused on using 3D scanning and printing to provide custom-fit medical products, and a web-based company that allows consumers to custom design their own household products. Along with developing an entrepreneurial concentration in the curriculum at Art Center, the timing of launching TDA should prove to be beneficial for the next generation of innovators. "In today's economy, artists and designers play a vital role as creative leaders and catalysts for innovation and change," says Art Center President Lorne M. Buchman. "Integrating technology, design and business, The Design Accelerator creates an opportunity for Art Center and Caltech alumni to expand on what they learned as students and provides a stimulating space for their concepts to become viable businesses."

THE FUTURE OF DESIGN AND TECHNOLOGY IN ACADEMIA: NEW MODELS/NEW SCHOOLS/ NEW PROGRAMS

The majority of this chapter has focused on the vital role partnerships and collaborations play in innovative approaches to design education. Academia is experiencing radical disruption, as this generation of learners is re-defining how they are accessing knowledge and beginning to embrace the idea of lifelong learning. Educational institutions are challenged with these disruptions as they face serious budget challenges, from funding for public schools to rising college tuition fees, along with student debt being at an all time high in the USA. Massive open online courses (MOOCs) are enabling children and adults to access free open-source world-class education in endless subjects. Charter schools are bringing the creative arts back into their programs and experimenting with new ways of teaching. Higher education is embracing the value of breaking down silos, and launching new multi-discipline programs and delivering more customized degrees.

One of the pioneers in democratizing education is Salman Khan, a graduate of MIT and Harvard Business School. In 2006 he launched the Kahn Academy, a non-profit organization providing a free world-class education for anyone anywhere. The online "school" features thousands of educational resources—from video lecture and tutorials (stored on YouTube) to personalized learning dashboards. Subjects range from math and sciences, to history and economics. The Khan Academy is currently delivering over 300,000,000 lessons with ~10,000,000 students per month accessing the extensive library of content. Students can participate in interactive challenges and assignments, finding mentors along the way. Parents and teachers have unprecedented visibility into what their students are learning and doing with the Khan Academy. This new global classroom is filled with millions of Khan Academy students from all over the world.

The Massachusetts Institute of Technology (MIT) has also pioneered a new open-access model with the MIT OpenCourseWare, a large-scale, web-based publication of MIT course materials. Educational materials from its undergraduate and graduate courses are posted online, partly free and openly available to anyone, anywhere. Over 2,000 MIT courses are available online, providing reading lists, discussion topics, homework problems, exams, and lecture notes. More and more courses are offering interactive web demonstrations and streaming video lectures, or can be downloaded for viewing offline through platforms like iTunes U. This bold initiative has prompted hundreds of other institutions to make their course materials available as open-educational resources through the OpenCourseWare Consortium.

Along with major access to online learning content, academia is witnessing an explosion of "Design+" programs, primarily in graduate programs, including dual degrees and new Design MBA programs.

The Stanford d.School was the first university program to create a hub for innovators—bringing together students and faculty from engineering, medicine, business, law, the humanities, sciences, and education—with a methodology for innovation that combines creative and analytical approaches. While this program is not an accredited design degree program, at the heart of its success is the human-centered design-thinking process, drawing on methods from engineering and design, and combining them with ideas from the arts, tools from the social sciences, and insights from the business world.

California College of Art (CCA) offers business degrees that focus on different aspects of innovation, integrating design and experiential thinking, systems thinking, sustainability, strategy, and leadership into their programs. The MBA in Design Strategy is ranked as one of the best business degrees for innovation in the world. Their most recent launch of the MBA in Public Policy Design and the MBA in Strategic Foresight programs addresses the emerging role design is playing in achieving innovation in governments, NGOs, and corporations.

Philadelphia University's Kanbar College of Design, Engineering, and Commerce (DEC) is the first in the US to bring together the three disciplines, offering a trans-disciplinary approach. The program provides students with the skills and knowledge to think creatively and to work collaboratively to discover innovative solutions to complex problems. The DEC curriculum emphasizes the value of collaborations between designers, engineers, and entrepreneurs, combining critical thinking and creativity with analytical thinking and business strategy, preparing students in solving today's real-world problems.

Art Center's latest graduate program, Innovation Systems Design (ISD), is a dual MS/MBA degree offered by Art Center College of Design and the Claremont Graduate University Drucker School of Management. The new program combines the Drucker School of Management's strategy, leadership, and management acumen with the rigorous development of creative skills and design innovation methodology found in Art Center's Graduate Industrial Design program, to best prepare tomorrow's innovation leaders. The program's mission of "Design Thinking and Design Doing" focuses on project-based education applying design-thinking methodology to business culture in order to create better futures. Students are immersed in Art Center's design culture for one year and then in Drucker School's business culture the following year, giving them the opportunity to apply their knowledge and skills through integrated project studios taught by working professionals, as well as faculty from both institutions.

In the last few years, the US Government, education policymakers, and school districts across the nation have been beating the drum on the value of STEM education, an acronym for the fields of study in the categories of science, technology, engineering, and mathematics. The idea of emphasizing STEM in schools from kindergarten through college is to improve the nation's competitiveness in technology development and to stimulate a thriving workforce. Another movement adding to the heated debate is STEAM, STEM + Art/Design. Entrepreneur and educational leader, Georgette Yakman, the initial publisher and founding researcher of STEAM, has found a formal way to link the subjects together and correspond them to the global socioeconomic world, defining the STEAM approach as: "Science and Technology, interpreted through Engineering and the Arts, all based in elements of Mathematics".

STEAM organizations, conferences and journals are informing the policymakers, and society at large, that real innovation is driven from all of these disciplines coming together to ensure a prosperous future. The Rhode Island School of Design (RISD) has championed this movement by encouraging the integration of art and design in K-20 education. The art and design in STEAM addresses creative problem-solving, translates complex data through visualization and brings innovative ideas to life. Former RISD President, John Maeda, participated in the *Stem + Art = STEAM* panel session at the 2013 World Economic Forum in Davos, Switzerland, stating the value art and design has in unleashing entrepreneurial innovation. INSEAD Chaired Professor of Innovation and Leadership, Hal B. Gregersen, moderated the discussion on the value of STEAM stating "If we get science, technology, engineering, and math folks into the same room with artists and designers and we have them actively do something—that is the best lubricant for the conversation."

From a designer and design education perspective, the future of meaningful learning in preparing the next generation of innovators requires a design-thinking, creative project-based learning approach. Not just in colleges and universities, but at all levels: kindergarten, primary school, and high school. The final example in this chapter comes from the Art Center Design-Based Learning Lab. Design-based learning engages students' innate curiosity by creating fun, interactive environments that develop higher-level reasoning skills in the context of the standard K-12 education. By leveraging design-thinking in developing curriculum, teachers can step away from the current traditional style of instructing, filled with one-way lectures and relying on test taking to evaluate students' learning. Design-based learning challenges students to create never-before-seen solutions to specific problems through visual communication and making. For example, a history class studying ancient Egypt, China, Greece, and Mesopotamia might begin by working in teams to build a 3D scale model of a place in that ancient world. Doing so compels them to answer questions such as "What types of dwellings will people live in?" and "What values will they live by?" Armed with their own solutions, students investigate topics about these societies in required texts. Instead of students sitting in class taking endless notes, dreading reading assignments, tests, and final exams, they are immersed, engaged, and excited about learning.

Design-based learning was developed by Doreen Nelson, a professor at Art Center College of Design and California State Polytechnic University, Pomona, and recipient of the California State University 2006 Wang Family Excellence Award in Education. In 2006, Nelson founded and currently directs The Summer Institute for Teachers, a five-day, intensive program taking place every summer in Art Center's Design-Based Learning Lab. This award winning program empowers teachers to tap students' natural creativity to develop higher-level thinking and enhance comprehension of the K-12 curriculum. In closing, a quote from one of the participating teachers: "Design-based learning will close achievement gaps and increase lifelong learning. I believe this would be the platform to truly reform and produce an educational system that meets the needs of all students. It truly taps into all learning modalities that help students access the core curriculum."

CONCLUSIONS

In this chapter we discussed and analyzed the dynamically evolving role of design education, especially in the light of technological discovery and progress. The discussions are based on many examples from the past, as well as the present, including an educated look to the future. The chapter describes and discusses in an over-arching and multi-disciplinary way how design is not only taught, but also how students get the best hands-on experience of what design represents in today's world and especially in today's business environment.

In particular, the following topics were presented, discussed and analyzed:

- From the beginning until today, a short historical perspective on design education. The discussion was structured into elements such as:
  - The vital role of partnerships
  - Access to collaborative processes and tools
  - Understanding the value of business and academia
  - The rise of social innovation
  - The entrepreneurial mindset
  - Design process inspiring education at large
  - The Millenial generation's approach.
- The story and examples are based largely around Art Center College of Design in Pasadena, California, but take many other examples and experiences into account.
- The "sponsored project" concept is presented and discussed in detail and how this close collaboration between academia, its faculty and students, and the industry led, and still leads, to re-defining products, brands, and systems, and especially customer and consumer experience.
- Design is discussed as a process for exploration and several examples, especially the Wonka Experience, together with other brands and companies are presented and analyzed.
- The concept of "DesignStorm™," Art Center's immersive innovation workshop is introduced and discussed, giving several relevant examples.
- Design creates relevance; it's not just an arbitrary style exercise, but defines the entire experience that surrounds a product or service. Again, several examples are discussed and analyzed.
- Another initiative at Art Center, called "Designmatters," and its practical relevance, is introduced and discussed on the basis of several examples.
- The idea that design is a process for transformation is strongly proposed and discussed with several examples of joint projects and partners, such as Caltech, Pasadena, California, or the United States Geological Survey in the Multi- Hazards Demonstration Project that was set up after the natural catastrophe that struck Haiti in January 2010.
- This chapter also discussed and analyzed the expanding role of design, from business being design driven towards design becoming business driven. It describes and analyzes in much detail the collaborative efforts between design colleges such as Art Center and business schools such as INSEAD, and especially emphasizes the critical role that "fault line discovery" plays, a discovery and innovation that takes place at the interface or fault lines between different disciplines.
- The chapter then discusses and analyzes the drive from multi-discipline to "über" multi-discipline design, and embarks on a discussion of the future of this "Design +",

where the "+" represents the fact that design alone is not enough and that it needs to be combined with other disciplines, not necessarily only from the art world, but technology, business or sciences, especially life sciences.

- Several areas of this "über-disciplinization" are discussed:
  - Integrating the "Cos" such as collaborate/co-create/co-locate/co-curricular
  - Design + technology + business = the DNA of innovation
  - Innovative spaces for learning.
- This chapter also discusses and analyzes ways how to prepare the next generation of innovators, which is largely based on the notion of the "experience portfolio." It introduces the notion of the "Design +" students that will come out of new and holistic ways of education and training, and which is closely linked to lifestyle and attitudes of the Millenial generation.
- The topic of venturing in design education and introducing a "Kickstarter" culture with crowdsourcing as the most normal way to create new value. It also discusses the trend found in academic organizations to go away from non-profit towards for-profit, thus instilling a sustainable business-driven culture from early on in the educational process.
- Finally, this chapter discusses the future of design and technology in academia by discussing and analyzing new educational models, but also the potential of new schools and new academic programs. Many examples are given and the outlook shows an immensely diverse picture.

TOPICS FOR FURTHER DISCUSSION

- Analyze and discuss your personal experience of design or technology education and its influence on the work environment. If no such influence is to be observed, discuss ways to change this and how you could potentially influence the process of change. Discuss this with your peers and bring the results to the attention of your management.
- Discover the role of design education in your direct environment, at school, in the workplace, in your personal environment, and attempt to find common elements, such as—only to give one or two examples—is part of daily activities, or plays an important role with management, or is non-existent in company-organized education, etc. Summarize the examples and discuss how to make the good elements work better and especially make design education part of the company education curriculum.
- How much evidence do you see in your work environment for design being a process for exploration, a process for creating relevance or design being a process of and for transformation?
  Collect all evidence and piece it together so that you get a good picture as to how this works in your organization, your company.
- In the process of putting together the above "evidence," look out for "fault line discovery" or the process of innovation at the interface of different disciplines. How much design and design thinking can you discover in this process? Analyze, synthesize, and discuss with your peers.
- This chapter talks about "Design +," meaning that teaching and learning about design in monolithic ways is no longer good enough in today's business environment.

- Do you see "Technology +", or "Science +" or "Logistics +" or "Marketing +" or "xxx +" applied and encouraged in your work environment. Please analyze, discuss, and especially look out for recurring elements that can make any discipline a "+" one.
- Analyze and discuss the future of design and design education seen from your personal and work perspective. Discuss in a group with participants who have all different work backgrounds and try to synthesize the major elements into a recurring pattern.
- Kindly let us know your results in a generic way.

REFERENCES

R. Dahl (1964). *Charlie and the Chocolate Factory*, Alfred A. Knopf, Inc. New York.

N. Howe and W. Strauss (2000). *Millenials Rising*, Vintage Books, New York.

5 Design and the business world

If you don't live out your own mad visions, leave the people who have a sense of humor alone.
Maira Kalman (The Art Center Design Conference, Pasadena, CA, March 2004)

DESIGN: THE HELPER FOR BUSINESS AND TECHNICAL

In today's consumer goods industry, and especially the food industry, it is widely accepted that design plays a secondary role, strictly executing what marketing people may have concocted and decided. And yes, there are many small examples and attempts in the industry that point to the contrary, but it appears that the importance and role of design in the food industry goes through cycles that follow each other.

When P&G set an example of design drive in the early 2000s (see Chapter 3), other companies followed, and from personal experience, as I was at the center of this evolution at the Nestlé company, I can say that it was not an easy task to convince management that it may have been a good idea to "do something" too. The "do something" was a large undertaking, but in the end resulted in the recognition of design being more than just a helper. But let us just stay a moment with the helper function, as this is still the prevailing situation. Without wanting to go into too much confidential detail, it can be said with certainty that the hierarchical level of designers (not design managers, who typically are not designers, but, as the title suggests, managers) in a food company such as Nestlé is at level 5 at best, rather level 6 out of a total of maybe 10 to 15, the latter very much depending on geography and organizational or structural set-up.

I am far from belittling the role or importance of hierarchical structures at any level. When working at the Nestlé head office, the ladies (yes, only ladies) that came to serve you coffee or tea in your office at 9 am and 3pm were extremely appreciated, and, without any hint of cynicism, really important, as they gave the days a structure beyond the ever ongoing meetings—a situation called "meetitis" or the sickness of having to go to meetings—and brought a very different world into the offices, friendliness and simplicity, easy going and down to earth. We will expand on the role and non-role of meetings in later chapters, as, especially in the context of creativity, innovation, and execution, they can be and often are rather counter-productive. At best, meetings can define who does what and when, at worst,

Food Industry Design, Technology and Innovation, First Edition.
Helmut Traitler, Birgit Coleman and Karen Hofmann.
© 2015 John Wiley & Sons, Inc. Published 2015 by John Wiley & Sons, Inc.

they just fill the days in the office. I mention this here in some detail, as these attitudes were and are the opposite of the regular corporate atmosphere, in which everything happened, including attributing roles to designers and technical personnel, for instance.

In this environment, design was often seen as the 2D culprit, sometimes also as the 2D opportunity. When I was in my first discussions on design with upper management, it became quickly clear that their definition of design was purely two-dimensional and focused on the graphic design of a package, primary or secondary, of any product. One can imagine the uphill battle that it took to convince management on the basis of internally unavailable examples that there is yet another, third dimension to design, not to mention any number of other design dimensions, the "n-dimensional space" of design, if we add media, movie, fashion, and other, earlier described design disciplines (Chapters 1, 2 and 4).

An almost subversive tactic to succeed is to create your own examples and success stories, creating standing for inventing examples or tweaking existing examples. It's a risky business, but often leads to success. As this book is also intended to give you some hints and suggestions on solving difficult situations, this comment has a well-deserved place here: invent your own, credible, hopefully do-able success stories and present them to management. Again, it's risky, but, in the absence of past examples, it's sometimes the only way.

Another, still slightly subversive, approach is to find examples from competitors. This is tricky, as it can sometimes be seen as condescending towards one's own organization, but more often than not, will succeed, as the comparison very often comes from top management itself. It is used as a kind of "agent provocateur," showing "our own people, how competition does it better than us."

If you have a helper status, use it smartly and to your advantage, and "beat them with their own weapons." This can simply mean that either you come up with the external examples first, hence even better prepare your arguments, proposals, or invitations to meetings. Anticipation is extremely important, which requires that you have a totally open state of mind when it comes to the best design solutions and don't get exclusively hung up on your own, great idea, which may, at the end of the day, still be the best solution. Management will greatly appreciate your preparedness in such situations; just make sure that you do not patronize, stay humble, and accept management's concerns and proposals: you do not necessarily have to accept them, provided you finish and execute with success.

From personal experience, this approach works more often than not. However, the best situation you can find yourself in is to be prepared with a list of great and realistic past examples from your own work, or work that you did in collaboration with others and that has relevance to the proposals that you are or will be making to management, to try to convince them to your side.

Another, often made observation is that management uses design for their own glory. This and other similar observations are not a rant against management, but should rather be seen as straightforward observations of real situations; they can happen as insinuated, they do not have to. One has to find out for oneself, depending on the environment one works, the kind of projects, the colleagues and experts involved, the management that decides on certain or many or all issues. When I say "glory," what I mean is that good design, as seen by management, reflects back on the designer and the decision-maker, and hence, on management that has ultimately taken certain design decisions.

It has to be remembered that we are still pretty much only touching the surface of what design really can represent to a company, especially a food company and where management still defines design pretty much two-dimensionally, although in some instances, improvements can be seen. In recent years, management has increasingly adopted three-dimensional thinking when it comes to design, yet the above-mentioned "n-dimensional space of design" is far from being understood. We will discuss and analyze this in more detail later in this chapter.

DESIGN: THE CONNECTOR OF BUSINESS ELEMENTS

In today's corporate reality, the various elements of the business are still very much discrete units, more or less collaborating well or often, not so well with each other. Many business and management books have been treating this topic, and "collaboration" is the topic of many management seminars, and consultants will also tell you how important it is to collaborate more closely with the other branches of your company. There is so much information out there on these issues that it is impossible to select the one book or article to reference here. Pick your own book from the book-shelves at airports or elsewhere.

When looking in a very simple and compact fashion at the value chain of any consumer goods company, especially food companies, there are the following branches, the "discrete elements" to be found:

- Research & development and manufacturing
- Logistics and supply chain (in-bound as well as out-bound)
- Commercial activities (marketing, communication, advertising and sales)
- Legal, finances, human resources, general administration
- Others that may be very specific to your organization (e.g. out-sourcing, in-sourcing, co-packing, store brands manufacturing, etc.). However, most of these will be found under the main four topics above.

When looking at the cost structure of these individual elements, it is very often difficult to attribute individual costs exactly, and they may vary from company to company, for instance depending on whether your company is a marketing and sales company without any internal R&D or even manufacturing, or whether you work in a food ingredients supplier company for whom R&D and manufacturing is the crucial and over-riding part.

However, it can be said that for large food companies that cater to retailers and consumers, R&D costs are typically between 0.5% and 2.0%, most likely on average 1.5% of the company's annual revenue, of course not counting manufacturing in this number. I put special emphasis here, because I want and need to make a few important points here. First, management's expectations are very high when it comes to the return on R&D investment/spend. Some food companies still see R&D expenses as "money spent," others regard it as "investment." My personal preference lies on the side of investment, although for companies who buy in R&D, it appears logical to define this as "spend." So, if expectations are high, pressure becomes high as well and scrutiny of results becomes even more pressing and detailed. It's this detailed interest of management in R&D results and also in the technical and design solutions that are proposed in any project that requires special attention by project and design managers, so that expectations can be properly and realistically managed. How often have we heard that a project was "over-sold", the enthusiastic, yet also naïve project managers had the natural desire to passionately talk about their results, the status of the project, and the great discoveries they had made. All that the managers heard, or rather wanted to hear, was that the results were so good that the project was ready to go and that, as of very, very soon, it would be in the factory, being manufactured and then sold with good profits. Two worlds clashed with each other, and the outcome was and is always very predictable: more R&D and discovery is still needed and the project is not ready for execution yet. "It's always the same," management say, "these R&D people are not capable of delivering, we always knew it and now, here is proof, once again."

The expectations of management are not only high, but often unrealistic, because the two groups, management, consisting typically of the kind of people that know everything (or a great deal) about the business, but have no or very little insight into the creative, innovative, and technical execution-driven process of research and technical development, ideally combined with optimal design so that the right products, ready to go into a factory and the marketplace, can be concocted and executed.

It is this absence of understanding that makes the dialog so difficult and sometimes impossible, as neither party takes enough time and effort to achieve a common language, a common ground. How can this be achieved? Again, from long-standing personal experience it's the designer's thinking and ways of achieving results. It's the designer's thought process that follows a different, form and function-based pathway, is a great connector, and can almost form the glue for these two very different worlds, that, at the end of the day, have or should have the same goal: to make better, new, more efficient, more profitable, and more consumer-preferred products for the good of the company, the other members in the value chain, and, ultimately, for the consumers.

To get to this common ground and better understanding requires quite a bit of effort on all sides. As can easily be anticipated, management, being at the higher hierarchical level, expects the product and design innovators, the technical actors, to come to their side, although it is clear that for a really successful rapprochement both parties have to come together. There is an old German proverb that goes approximately like this: "The smarter one gives in." I let you judge who this should be.

THE "N-DIMENSIONAL DESIGN SPACE" IN THE BUSINESS ENVIRONMENT

Earlier on, we mentioned the "n-dimensional design space." Let us discuss this in more detail and look at the impact and consequences this may have on the business side of a consumer goods company, using examples of food companies. In mathematics, n-dimensional space describes a situation in which the thought process moves out of the static, Cartesian mathematical paradigm to a dynamic one, in which more than the usual, imaginable three dimensions exist. The fourth dimension is accepted to be time, a temporal dimension, whereas further, fifth, sixth, nth dimensions are again considered to be spatial and mostly necessary, typically used in Superstring theory and Bosonic string theory, just to name a few.

However, we use this example of n-dimensional space not for its beauty and enigmatic characteristics in mathematics and physics, but to better describe what we mean by "n-dimensional design space" and what it may represent in a business environment such as the consumer goods industry.

We use the comparison on two levels. The first is straightforward and relates to the large number of design disciplines as already briefly discussed in Chapters 1, 2 and 4 and reinforced in this present discussion.

Business as well as designers have to accept that in theory as well as in practice there are n numbers of design disciplines, even those that have not been invented yet and go beyond the existing, imaginable and trivial ones. In other words, everything is design and, in principle, every problem can be resolved by applying any type of design and especially design thinking in either simple or complex ways. It is the real n-dimensional space of design, in which "design thinking" becomes the fourth dimension of design and can be seen as either "temporal" or "spatial," but mainly just pragmatic and common-sense driven. These terms do not exist in the context of design yet and we introduce them here for illustrative purposes and better distinction.

"Temporal" design describes the situation where design clearly understands its environment and the target consumer groups for which it designs in the largest sense. This is not always easy to achieve, as it requires a deep knowledge and understanding, plus lots of anticipation and guts to do good design in the time frame that it can work and for the consumer groups for which this design is intended. It is temporal, because it does not survive long periods of time; it is finite and evolves at a fast pace. Typical examples are objects that have been designed in areas such as automotive, consumer electronics, or appliances. What was considered good and modern design this year will have fallen out of fashion next year, although certain underlying design trends certainly last longer. An example of the latter is, for instance, the rather "edgy" and straight lines design of automobiles in the 1970s and 1980s, whereas today the underlying theme is definitely curvy and round, as, by the way, it was in the 1950s and 1960s before everything went to straight lines. It is to be expected that the straight lines will come back again, it's a matter of time, maybe another five or ten years. Figures 5.1 and 5.2 illustrate this back and forth of style.

Figure 5.1 VW Golf 1; around 1974. (Reproduced with kind permission of AMAG Automobil- und Motoren AG, CH-5116 Schinznach-Bad, Switzerland).

Figure 5.2 VW Golf 7; around 2013. Reproduced with kind permission of AMAG Automobil- und Motoren AG, CH-5116 Schinznach-Bad, Switzerland).

"Spatial" design, on the other hand, describes a situation in which the designed object is "timeless," has been created many years ago, and has survived the changing times. The food industry, especially, has many examples of what we call "spatial" design. Confectionery products such as Nestlé's and Hershey's KitKat® or Mars' SNICKERS® Brand and other similar icons of the food industry were introduced many years ago, some of them over 70 years ago, and their general appearance and design are almost unchanged, as illustrated in Figures 5.3 to 5.6. I define them as examples of "spatial" design because they have kept their space over extremely long periods of time, have gained fame and notoriety and solidly exemplify great and timeless design.

It is not the subject of this book to discover and propose new, yet undiscovered design disciplines, but we will clearly focus on the smart combination of the existing and most relevant disciplines of design in combination with design thinking, which inevitably again leads us to the need of establishing design right at the decision center of a consumer goods company, especially a food company.

Designers and other members of the business and technical community have to start thinking in multi-level terms, i.e. looking beyond their immediate environment of expertise, and willingly and with great enthusiasm connect to their colleagues "across the border." Many do not dare to do this, as they might be afraid to expose themselves to areas unknown to them and uncharted territories with potential new and surprising discoveries around every corner. And yet, it is the only truly successful approach to real and sustainable success in business. Without the connectors and the approach they propose and live, "numbers can be made" at the end of the business year, but in the medium and long term, such business will dry up and not be successful achieving these results year in, year out.

Figure 5.3 Picture of Historic KitKat®. (Reproduced with kind permission of Nestlé SA, Vevey, Switzerland. Nestlé and KitKat are registered trademarks of Société des Produits Nestlé S.A., Vevey, Switzerland).

Figure 5.4 Picture of Contemporary KitKat® Logo. (Reproduced with kind permission of Nestlé SA, Vevey, Switzerland. Nestlé and KitKat are registered trademarks of Société des Produits Nestlé S.A., Vevey, Switzerland).

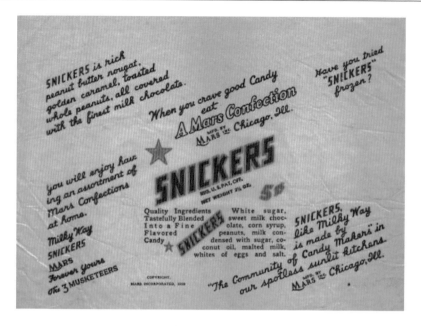

Figure 5.5 Picture of Historic SNICKERS®Bar Packaging. (Reproduced with kind permission of Mars Chocolate, North America, Hackettstown, NJ, USA).

Figure 5.6 Picture of Contemporary SNICKERS®Bar. (Reproduced with kind permission of Mars Chocolate, North America, Hackettstown, NJ, USA).

TYPICAL AND DESIRABLE BUSINESS INTERACTIONS INSIDE TODAY'S CONSUMER GOODS INDUSTRY

Much is talked about the necessity of good interaction and collaboration between the different branches, departments, or business sectors of the industry. This is certainly a very important aspect and often this is really lived and well applied in many ways. However, the reality behind this situation often presents itself in a very different fashion: isolationist, jealous, keeping information to myself, not communicating enough, singly hanging on to successes, and one could find many similar descriptions of such situations. It's not always nice in business, one could even say, it often is ugly and borderline childish.

So, what's going on in the consumer goods world of business? In this section we do not want to put too much emphasis on what's not working or list examples of bad behavior and unprofessional ways of collaboration, but we will attempt to illustrate just a few, very typical situations that paint the picture.

Top management asks for more detailed information on the status of a project, without telling the project manager why they want to have this information. What then happens there is the "overkill" of preparing all the information; everything that could be relevant, and often totally irrelevant, is then compiled, and detailed presentations are prepared. If, on top of the request for project-related information, top management announces a visit to the project team's location, the whole situation becomes even more tricky, and especially time and money consuming. The "full Monty" is put in motion and in addition to the preparation of detailed presentation material, rehearsals are the order of the day. The project members and middle management spend many hours together just to make sure that every possible situation is simulated and every possible question is asked, and relevant and hopefully satisfying answers are given. This is also a busy time for graphics designers; they really are put to work to prepare trendy slides and colorful posters.

Often the technical experts will call their marketing colleagues—definitely a worthwhile undertaking—to ask for their business input, mainly to choose the right words, the right "speak" or "lingo" that hopefully will make top management tick. I could continue this story with details like management being late for the meeting so it is cut very short, management asking very different questions from the ones the project team has prepared, management all of a sudden wanting to talk about another project that in the meantime has become more important, and so on and so on ... Many of you must have encountered similar situations in your professional environment, from either side, and must have asked yourselves questions about the sense and nonsense of such situations. You may have even calculated, just for curiosity, the amount of time put into such an event and, including evening and weekend sessions and the number of people involved, you may have come to outrageous, possibly five-digit numbers.

It is, however, interesting to observe that, despite negative and thoroughly realistic examples, the industry, or at least many important players, really do well and are successful. Does this success come despite the mainly head-office-related inefficiencies or are these inefficiencies a stimulus to the other branches of the company, a kind of "let-us-show-you" attitude? Although it is almost impossible to prove, yet based on many personal observations, it is suggested that business could be doing even better were they only to adopt the simple and easy to apply rules of mutual respect, and close and efficient collaboration.

So what might the ideal interaction look like? First, we have to look into the number of hierarchical layers, professional structures, and other topics, such as dual career ladders (commercial and technical), and, not least of all, personal attitudes and behavior, the latter being largely driven by career opportunities (the "carrot"), as well as financial outlook. Although it is often said that money is not a driver for performance and efficiency, the converse seems to be true, at least in the majority of people that mingle in the system. Money is, for instance, compared to a piece of soap and the need to wash dirty hands: "as long as I have a piece of soap with which I can wash my dirty hands, it is irrelevant how big the piece is," and that should be true for money as well: as long as I have enough to live, sustain my family and be happy, I do not need more.

I do not know who came up with this image, it's a nice one by the way, but it simply does not reflect the reality. Reality is more complex and managers have to understand that money is one important driver in a mix of a few others of probably equal importance, such as freedom to act, responsibility for the work I do and my environment, successes singly or jointly achieved and, last but not least, having fun (or if you prefer pleasure) in doing what I do! For the business to run properly and even better, it is really important that management and personnel recognize the importance of each of these aspects, and possibly others not listed here, but of relevance for your specific organization.

Another "feel-good," soft-wash slogan is the one that all of you must have heard many times over: "People are our most important assets." It is supposed to demonstrate that management really cares about you, because without you, things just simply wouldn't work. It is a truism, but it does not reflect the reality; I was almost tempted to say the "real reality"—it's not just people that are the most important assets, it's the RIGHT people. This is maybe tough to swallow, as it puts a lot of doubt on yourself, whether you are amongst the chosen or perhaps do not belong to this group. I shall discuss this concept of "the right people" in much more detail in Chapter 9. Responsibility sits with both sides, management and employees, to be honest and straightforward in the assessment of the "right ones." Without such honesty and subsequent consequences, no good collaboration can be achieved between different branches or departments of a corporation and the different layers and levels within the organization.

We could go on with more examples of this, and how to optimally collaborate within any organization, small or large, but I think the above gives probably the most important examples and hints as to what might be of greater importance to the wellbeing and success of your organization. It's just a side topic of this book, yet I believe a crucial and important one, because all of what we present and discuss in this book, in any chapter, is based on people, their actions, and how they ideally and efficiently interact and collaborate to achieve the successes they jointly strive for.

DESIGN: THE ENABLER FOR LOGISTICS AND SUPPLY CHAIN

Like many members of a corporation, and especially in a food company, designers and appropriate design can and must play an important role in the actions of their colleagues in logistics and supply chain. Let us first take a very top view glimpse into the activities of logistics and especially the supply chain.

The main building blocks of logistics and supply chain in the food industry are complex, yet rather straightforward. Typically, the procurement department looks into these aspects on the "inbound" logistics side, the inbound supply chain. These activities partly take place inside the organization and are partly outsourced. The various, very simplified building blocks on the inbound side are:

- Procurement of raw and packaging materials and services, including supplier selections, preferred suppliers, negotiations of prices, establishment of term sheets (in collaboration with legal department), other tasks.
- Organization of inbound material and services flow, including transportation; the latter is most often outsourced to specialized transportation companies, mostly done by suppliers.
- Ideally, early participation in project and design work; we will elaborate on this point in more detail in Chapter 8. This is the most natural connection between logistics/procurement and the activities of the development people, and especially those of the designers.

The outbound logistics, the outgoing supply chain, is possibly more complex, as it is, from personal experience, not as centrally co-ordinated as the inbound side.Let me give one typical example: whereas the procurement of packaging materials is often centralized and grouped together to achieve economy of scale and optimal, i.e. low prices, and whilst this can be and most often is done by a central group for the entire operational geography of a company, manufacturing of products is always done in one specific geographic location. Transportation and distribution of such manufactured goods is organized locally, ultimately in a great number of different locations, very often not really over-seeable and not necessarily always under the best conditions, both economically and with regard to storage conditions.

Both in the area of inbound and outbound supply chains, good design and technical smartness can add a lot to improve the conditions under which raw and packaging materials come into the factory and especially when it comes to optimally storing, shipping, distributing, and selling food products of any kind.

Let us illustrate this with the example of manufacturing a chocolate wafer product such as KitKat, or a Mars Bar or any other similar product. Raw materials as well as packaging materials have to be ordered and procured under the best conditions and the most appropriate suppliers (raw materials, packaging materials, converters/printers, possibly others) have to be selected. It may be assumed that all suppliers have been well known to the company for quite some time already; nevertheless, it is important at all times to scan the market and be vigilant so as to always deal with the most efficient supplier at all times. But how do we know the most efficient supplier and by which criteria do we measure? Hard-core purchasing experts will definitely go for best price and will push for online auctions, especially reverse auctions.

Definition Definition of Reverse Auction from "Investopedia" *http://www.investopedia. com/terms/r/reverse-auction.asp*: A type of auction in which sellers bid for the prices at which they are willing to sell their goods and services. In a regular auction, a seller puts up an item and buyers place bids until the close of the auction, at which time the item goes to the highest bidder. In a reverse auction, the buyer puts up a request for a required good or service. Sellers then place bids for the amount they are willing to be paid for the good or service, and at the end of the auction the seller with the lowest amount wins. Reverse auctions gained popularity with the emergence of Internet-based online auction tools. Today, reverse auctions are used by large corporations to purchase raw materials, supplies and services like accounting and customer service. It is important to note that reverse auctions do not work for every good or service. Goods and services that can be provided by only a few sellers cannot be acquired by a reverse auction. In other words, reverse auctions work only when there are many sellers who offer similar goods and services.

From the definition given above, it is clear that reverse auctions are not the universal solution for all procurement challenges, although there is also much to be said for them. However, it is important to apply this approach with a grain of salt, or better with a whole heap of grains. In an organization such as a food company, it is important that commercial, technical, and design savvy experts work together to approach this with the necessary smartness, as well as caution. Most importantly, there have to be weighted criteria that are put in place to appropriately judge the various important factors that are of relevance when purchasing raw or packaging materials. Such factors typically are: price; technical, design, and

R&D capabilities of the supplier; assured uninterrupted delivery of the materials; mutual respect (cultural fit); geography; quality and safety of materials, in the supply chain as well; possibly other criteria of importance in your own company. Now, how are these to be weighted? The answer to this question might only be given after a "clash of the titans" within your own company, with the most important point here being, don't leave the field to procurement alone. Technical experts, designers, and commercial experts have to discuss this in depth, and a good starting point is always, whether a packaging solution, for example, is going to be a "stock" solution or a "custom," bespoke solution. This gives us a first hint as to how to weight price in this whole balance of criteria.

The most important criterion, above all others, is assured quality and safety of the raw and packaging materials. If this is not a given, don't deal with the supplier. Once this criterion off the table, then you may go on to discuss and weight all the other criteria. From my own personal experience, it can be said that for stock items, and that also counts for stock ingredients, price weighs at least 50%, often as much as 70%, i.e. and alas, it is the major, decisive driver, probably followed by an assured uninterrupted supply. For custom ingredients or packaging solutions, criteria such as technical capabilities can often become the over-riding ones, and also cultural fit may play an important role here, as well as geography, thus pushing the price rather further down the ladder, maybe as low as just 25%. As you can see, the ultimate weighting of criteria is not at all an exact science and will always be influenced by many considerations, with many players having to have their say.

Finally, we have all the materials in the warehouse of the factory and manufacturing of the product can begin. Technical packaging experts, as well as designers, have to put their heads together to find the best technical and form solution for the manufacture of the specific product, to achieve the fewest line interruptions (mean time between failure or stoppage), the easiest collection of individual items at the end of the line to feed into, for instance, a horizontal flow wrapper that may run at speeds of up to 1000 or more a minute (typically probably less, maybe 400 to 600 a minute). It is easy to see that in each of these steps, the product is prone to breakage, thus becoming unsellable. Products that are abused in such ways within the factory's production lines can be re-used as "re-work." However, there is no average percentage of rework in confectionery factories to be found. Such a number is very dependent on the type of product; for instance, a line that produces only milk chocolate and has no changeovers to white or dark chocolate will inevitably have less rework, of the order of 0.5%, than a line with many changeovers to many different types of ingredients, which might run at approx. 3% of rework (L. Baumberger, Personal Communication, June 2013). "The story goes that production in one factory fiddled the re-work figures so much that there was more product leaving the factory than there were ingredients going in!" (S. Beckett, Personal Communication, May 2013).

It has to be said, however, that confectionery factories are masters at using re-work for making new, bespoke products. The "Cailler Branche" from Nestlé, shown in figure 5.7, is a typical example.

Yet not all re-work can be used in such clever ways and will ultimately go to other uses, such as the "doddings," as they are lovingly called by the production people, little spheres, typically over-baked from the edges of the baking plates that are an unavoidable by-product of the wafer baking process as it is presently done. Therefore, there is a fine balance to be respected between re-usable re-work that goes into the production of a food product and by-products from the manufacturing process, or re-work that cannot be used any more internally. This delicate balance between manufacturing optimization, product strength and integrity, reduction of re-work in all steps of the outbound supply chain, and need for re-work that is related to the creation of other products cannot be abdicated to

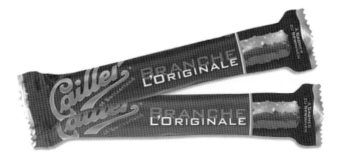

Figure 5.7 Cailler Branche "L'ORIGINALE" ® from Nestlé. (Reproduced with kind permission of Nestlé SA, Vevey, Switzerland. Nestlé and Cailler are registered trademarks of Société des Produits Nestlé S.A., Vevey, Switzerland).

manufacturing experts alone, or supply chain experts alone, and it is stressed here, once more, that the role of the designer is a crucial one, by bridging all these different steps.

Finally, it must be emphasized, and it is a very obvious fact, that once the product is in its wrapper, any breakage during distribution and warehousing leads to what is defined as "bad goods." Such bad goods are basically unusable and the retailer will typically send them back to the manufacturer, thus reducing the product's margin in sometimes non-negligible ways. A typical industry standard, especially in the confectionery industry, specifies bad goods to be of the order of 3%. This means that for every 1,000 tons of product, 30 tons go back to the manufacturer, or, for products such as KitKat or Snickers that have worldwide sales of far beyond 100,000 tons, this results in at least 3,000 tons, more likely 5,000 tons of bad goods, which may amount to 50 or more millions of dollars of lost sales. Therefore, one of the primary targets for anyone working in the supply chain is to strive for a reduction of bad goods; it can easily be calculated that reducing bad goods by just 1% could increase the sales value by around $20m or more, calculated very conservatively. Appropriate packaging with clever packaging design and efficient and careful, stress-free, and temperature-optimized product distribution are the main criteria to achieve the crucially needed bad goods reduction.

DESIGN AS A COUNTERFEIT FIGHTER

Counterfeiting is a well-known problem in any consumer goods industry and the food industry is not spared from this very distracting, margin-reducing, and often outright dangerous criminal activity. In addition, counterfeiting can damage the brand image and destroy consumer confidence in any brand. Counterfeiting is common in industries such as watches, cloths, shoes, bags, and T-shirts, as well as entertainment. Both music and movie DVDs are heavily counterfeited, although police apparently have really cracked down on this in South East Asia. Wandering through Petaling Street in Kuala Lumpur, Malaysia, gives you a good impression what's going on today. Watches are probably still the most prominent candidates for counterfeiting, but the stalls are filled with bags, shoes, small and large leather bags, and T-shirts in all sizes, including for Western, often heavy-built men, for whom the size is proudly called "King Kong size."

Never have I seen counterfeit food products being sold in such a market, which, of course, does not mean that it could not happen. But it happens more sophisticatedly, more discreetly, more in the dark, first food products and packaging materials illegally being filtered out of the supply chain, creatively transformed into new, almost identical-looking packaged products and then sold in different, often more shady retail environments. It even happened, and I have personally experienced this, that a new brand of soluble coffee was created by Russian counterfeiters. It was a very creative counterfeit, but it was not the real deal.

One can argue that, as long as there are no health and safety issues involved, the best thing may be to not even talk about it much and, behind the scenes, through and with the help of trade networks, slowly but steadily stop such counterfeit activity. And this is certainly a strategy that pays off over time; however, it will always unwittingly allow for some counterfeiting to happen.

This becomes much more critical where more nutritional food products, especially those for babies or small infants, are concerned. It could become outright dangerous. Counterfeiters want easy and quick money, they most of the time are not concerned with health and safety issues, and the manufacturer has to be increasingly, better totally, vigilant when it comes to a product such as infant formula. Too many bad examples could be read about, especially with ingredients stemming from unknown, uncontrolled, and unsuitable suppliers and supply chains. It has become very difficult for the consumer to detect counterfeit products, as some of them may even enter mainstream supply chains and mainstream retailers. The following is a short article from *Malaya Business Insight* in the Philippines from January 2013, reporting on a counterfeit Maggi product:

Fake Maggi products seized

Details Published on Monday, 28 January 2013 23:00

THE PNP-Criminal Investigation and Detection Group-Anti Fraud and Commercial Crimes Division recently seized fake Maggi Magic Sarap products in Quezon, Laguna, Cavite and Batangas. The fake items were seized in 77 retail establishments in public markets in Biñan, San Pedro and Calamba City in Laguna; Dasmariñas City, Silang, Carmona, Imus, Trece Martires, Kawit, Naic and Bacoor in Cavite; Tanauan, Malvar, Lipa City, Ibaan, Lemery and Bauan in Batangas; and Sariaya, Tayabas, Lopez, Gumaca and Lucena City in Quezon Province. The establishments were found to have been buying fake Maggi Magic Sarap from illegal operators at low prices. These retail outlets in public markets were then sold the fake products to consumers at regular retail prices. Fake Maggi Magic Sarap does not have the taste profile preferred by consumers and does not meet the rigorous quality standards of the genuine brand. To protect consumers, Nestlé requests those who have been misled by these counterfeit products, or know of others with such experiences, to report incidents to the Nestlé Consumer Services at hotline 898-0061 (Metro Manila) and 1-800-637853 (provincial areas).

Although such products are not in a high price range, such as luxury watches, where counterfeits are a factor of 100 to 500 cheaper, they are apparently still very attractive to counterfeiters. However, when looking at just a few recent headlines regarding counterfeit foods, one clearly sees the enormity of the problem. The following are just a few, very short articles shedding some light on the size of the opportunity for the counterfeiters and the size of the problem for the food industry. (http://www.havocscope.com/tag/counterfeit-foods/)

Amount of Food Sold in Italy in 2010 That Was Mislabeled

in COUNTERFEIT GOODS

In 2010, up to $1.4 Billion (1.1 Billion Euros) worth of counterfeit foods were sold in Italy, according to a think-tank based in the country. Counterfeited items included fake Parmesan cheese and spaghetti.

The total counterfeit goods market in Italy was worth $9 Billion in 2010, causing a loss of $2.2 Billion (1.7 Billion Euros) in tax revenue.

(Source: Ciancio, 2012.)

Types of Mislabeled Foods in the United States

in COUNTERFEIT GOODS

According to a study by the Anti-Counterfeiting and Product Protection Program at Michigan State University, the following food items were found to have been counterfeited.

The study analyzed its product database of counterfeit items, and found that 16 percent of counterfeit foods involved olive oil, 14 percent involved watered down milk, 7 percent was counterfeit honey, and 2 to 4 percent of the counterfeit items were fruit juices.

Worldwide, counterfeit foods create a $49 Billion market.

(Source: CBS Miami, 2012.)

Italian Olive Oil Made with Substandard Ingredients

in COUNTERFEIT GOODS, TRANSNATIONAL CRIME

Due to the role of organized crime in food production in Italy, a report has found that 80 percent of the olive oil produced in Italy and stamped with a "Made in Italy" logo was made with cheaper, lower quality oils from other countries.

Source: Leslie Clarula Taylor, "Italian olive oil part of organized crime probe," The Star, January 27, 2012.

Looking at the size of the worldwide counterfeit foods market, as described in one of the articles above, we can see that this amounts to roughly half of the annual revenue of a company the size of Nestlé.

In other words, it is really big, and hence of real interest to the criminal minds that inhabit the industry.

More recently, much of the counterfeit activity is now happening online over the Internet.

Richemont, an important player in the Swiss watch industry has just launched a program to shut down thousands of relevant sites (2,700 to be precise), as well as 4,900 domain names, such as "Fakeswatches.com" or "Replicawatchshop.us" (*LeMatin Dimanche*, 2013). It is clear that the Internet is and will increasingly become an important forum for counterfeit activities. The really important question is of course, how can and do we deal with this issue?

As briefly mentioned earlier in this section, one alternative is to do nothing of any spectacular nature, like seizing in big style and with much media attention, and strongly publicly condemn the actions of counterfeiters. Yet, in the background be very active in a kind of intelligence operation by going up the supply channels to the ingredient suppliers, packaging suppliers, machine suppliers, and distribution companies, and attempt to identify the counterfeit channels, almost in a covert action, and eventually dry them out, like a river that is cut off from its water supply. This is an approach that is rather effective and is consistently done within the industry. However, there are probably even more efficient ways to tackle this problem and outsmart the counterfeiters.

Basically there are three main approaches: speed, speed, and more speed. The industry has to be faster at developing new products and renovating existing ones, have designers and technical experts working together in the fastest possible way, have marketing to agree

Figure 5.8   Panoply of ice cream novelties from the Nestlé Company. (Reproduced with kind permission of Nestlé SA, Vevey, Switzerland. Nestlé and depicted ice cream brands are registered trademarks of Société des Produits Nestlé S.A., Vevey, Switzerland).

on new or renovated product solutions in a speedy manner (avoid the "ping-pong" between technical and marketing, where the technical solution proposed is never a good one and it is repeatedly sent back by marketing to make it a little more red, or bigger, or square, or cheaper, or, or, or …). In simple words: always be the first one in the marketplace with renovations or innovations, and renew frequently, so that counterfeiters cannot really follow anymore. From personal experience I have observed that ice cream products seem far less counterfeited as other food products such as dairy, infant formula, soluble coffee, and packaging material in general. Ice cream is a "novelty" business, a business that changes the content all the time, but more so the look of its products, ideally every three to six months, and therefore is much more difficult to copy with counterfeit products.

So, "It's the speed, stupid!," some former US president would have said, had he worked in the food industry.

The set of images in Figure 5.8 gives you a very illustrative picture, albeit rather confusing, of the variety and complexity of ice cream products.

Figure 5.8, admittedly being rather Nestlé centric, would not look any different if the brands were from Unilever, Mars, or any other big or small industrial ice-cream maker. You can easily understand the difficulty in timely counterfeiting of such a large variety of products, hence making complexity and speed the two biggest deterrents for counterfeiters.

Once more, the close collaboration between technical product and packaging experts, and the commercial and marketing representatives, well knit together by the work of designers, is an absolute must in successfully fighting counterfeiting and, if not bringing it to a halt, at least reducing it to insignificant levels.

THE WAY FORWARD: "DOWN-TO-EARTH DESIGN"

How can design and designers make their marks in the business world if their design solutions are too complex, too difficult to understand, and too controversial to be accepted by business people? The answer is simple: "Make it simple stupid" (MISS), a variation of the well-known KISS theme, basically meaning the same, but in a more active way. This is not a new concept, by any means, and you must have heard this time and again, perhaps cannot even hear it anymore. But, has it been applied in your environment, not only in the field of design but in any technical and commercial area of the food industry?

General managers may argue that they follow this principle, mainly for two reasons: first, they don't have time to understand and deal with complex topics and second, they may not want to deal with something complicated, following the slogan of "reducing complexity." The latter is intuitively right, but we will discuss the value of complexity a bit further down, when we discuss the complexity of processes and how we can profitably make use of the art of managing complexity. When it comes to technical and design matters, however, it must be said with a strong voice that simplicity trumps all!

Here's what Jonathan Ives from the Apple Company has to say about simplicity:

> *Simplicity is not the absence of clutter, that's a consequence of simplicity. Simplicity is some-how essentially describing the purpose and place of an object and product. The absence of clutter is just a clutter-free product. That's not simple.*
>
> (*The Telegraph*, S. Richmond, London, May 23rd 2012.)

In the title of this section I called it "down-to-earth-design" (DTE-design). This is not really a new discipline of design, but rather discipline within design. By almost religiously following the earlier described principle for good design, namely the balance between form and function, DTE-design is an automatic consequence, yet, it is by far not the norm yet, and the food industry is no stranger to this.

I have personally seen young product managers being given the task of renovating not so successful tier 2 or tier 3 products. Typically, their first reaction was to look at the graphic imagery of the packaging, only to decide that an additional color or a fancier font for some of the printing on the wrapper would be the best solution. Nothing is further from the truth: success is not achieved by making the wrapper MORE colorful and the font MORE fancy, but rather the contrary would be needed: be rather minimalist, FEWER colors and LESS fancy print would probably create a more desirable product look overall and thus enhance sales and bring the tier 2 product back onto the track of profitability. It goes without saying that this is not a universally valid example and there may be situations where the addition of a color to the candy bar wrapper can make it more desirable and ultimately successful, but these are more the exception.

It has to be noted that junior (or senior) product managers very often do not consult a designer, but instead discuss with the print experts in the converter's factory. Such an expert may dispose of a large, 11-head printing machine on which he may currently only use 5 or 6 heads, and therefore has empty "real estate" on his machine for sale. The interest is clearly skewed towards the technical capability rather than good, DTE design, with simplicity, and form and function, and ultimately the consumer in mind. But DTE design is also not necessarily wished for by designers. No wonder, as without close collaboration between technical, marketing, and design, the real consumer intentions and likings cannot be identified. It's yet another example, in which a fine balance between consumer pull and technology push has to be achieved.

DTE design is a concept that will still have to be understood by all players in this field, and here especially the drive towards simplicity and minimalistic solutions is often seen as a lack of imagination and creativity; and yet, the opposite is true! It's pretty much the same as writing short letters or notes that may take more time than writing something more elaborate, driven by the desire to be as complete as possible, maybe full of unnecessary details, and the inability to choose between what is absolutely necessary and what is superfluous.

The following illustrates a few examples of true, minimalistic DTE design to better illustrate this concept. One good example from the furniture industry is the Jørgen Møller Taburet   M   (https://www.google.ch/search?q=jorgen+moller+taburet+m+stool&tbm=isch &tbo=u&source=univ&sa=X&ei=Te6xUufoKYXe2AWhvYCYBg&ved=0CD0QsAQ& biw=1431&bih=740)

Although an extreme translation of the combination of simplicity, and form and function, here more functionality, it really demonstrates what DTE design can deliver: a great combination of design, style, and usefulness, all combined in one simple and straightforward item. Or just look at a Coke can: simple, eye-catching, cold, real DTE design. The Google Logo is another great example. Total simplicity results in the highest degree of recognition. The Nespresso® capsules, shown in Figure 5.9 are another expression of DTE design.

To conclude, two examples from the food industry that are not solely based on packaging design, but rather straightforward product design. The first example is based on the transparent package to show the various layered contents of the yogurt product. One can argue about liking or disliking the design itself, but it clearly shows the drive of the food industry towards better design that goes beyond packaging, or marketing and advertising, and that embraces the principle of simplicity and the DTE design concept.

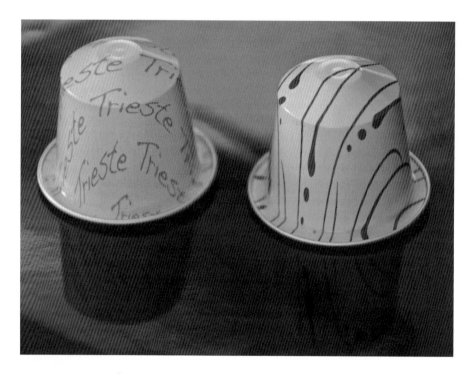

Figure 5.9 Nespresso® 2013 Special Edition Capsules—trendy and timelessly modern. (Image copyright of H. Traitler. Used with kind permission of Nestlé SA, Vevey, Switzerland. Nestlé and Nespresso are registered trademarks of Société des Produits Nestlé S.A., Vevey, Switzerland).

Figure 5.10 Cream-enriched yogurts (®"Migros Séléction"), Migros, Switzerland, June 2013. (Reproduced with kind permission of Migros-Genossenschafts-Bund, CH-8031 Zürich, Switzerland).

Figure 5.11 shows the product "Taro Chips," chips made from the taro root. You may judge for yourself the really pretty design of this product. Best of all, nature designs it this way!

We could add many more of similar examples, but the message is always the same: the simpler and more minimalistic the design, the higher the chances that such design is remembered and gains the all-important familiarity in a large consumer base. This is what DTE design is all about: being easily remembered and recognizable by the consumer!

THE FUTURE: DESIGN IS MANAGEMENT

In the introduction to this book, at the very beginning of Chapter 1, it was proposed that design thinking and designers should become a part of the general management, the executive board of any company, especially food companies. We briefly discussed the term design thinking in Chapter 3 through, the example of R. Martin's book on "The Design of Business" (Martin, 2009), but let us add yet another definition here to make this term a little bit more tangible.

Wikipedia proposes the following definition, which, in my eyes is a bit complex, yet rather complete:

> As a style of thinking, design thinking is generally considered the ability to combine **empathy** for the context of a problem, **creativity** in the generation of insights and solutions, and **rationality** to analyze and fit solutions to the context. While design thinking has become part of the popular lexicon in contemporary design and engineering practice, as well as business and management, its broader use in describing a particular style of creative thinking-in-action is having an increasing influence on twenty-first century education across disciplines. In this respect, it is similar to systems thinking in naming a particular approach to understanding and solving problems.
>
> (http://en.wikipedia.org/wiki/Design_thinking, April 2013)

The balanced mix of empathy, creativity, and rationality is a very good descriptor of what is meant by design thinking and would immediately suggest that the mix is not carved

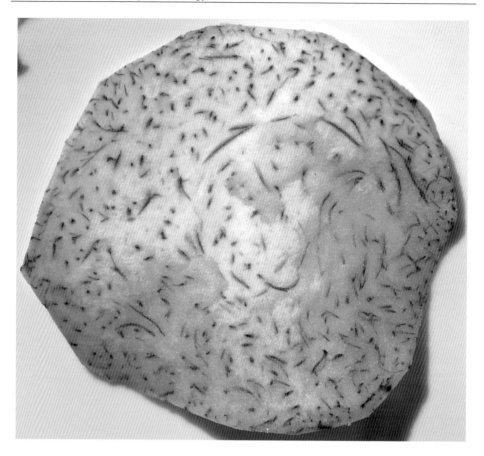

Figure 5.11 Taro Chip. (Copyright of Image: H. Traitler, January 2014).

in stone as to the individual weighting of the elements, but that their relative importance may vary with the situation.

Why do we propose this here so strongly? Given all the examples regarding the importance of design in many consumer goods industries, and especially the food industry, it would be a stark error to ignore the important role that design plays in all aspects and levels of the industry in general. Bringing the customer (retailer) as well as the consumer into the equation, this sense of design urgency becomes even stronger and the real need for not only design, but understanding design and designers properly becomes almost a centerpiece of the discussion here. Time and again I have seen managers and especially top managers of food companies making design-related decisions without any real and proven background, just on an intuitive (not so bad) or, more often, just a hierarchical (very bad!) basis, and thereby endangering the success of the brand for which such a decision was taken. Luckily for them, in the past, good and functional design was not necessarily at the center of customers' and consumers' requirements, but this has changed since and the changes are still ongoing.

Many factors play a role here, such as increased drive towards sustainable solutions, desire to properly separate and recycle materials, consumers' increased desire for good-looking products, which started in the automotive industry, conquered the consumer electronics industry, and finally has also taken root in the food industry. Amateurism is no longer an option; hence the forceful request to install design thinking and designers in person in the ranks of the executive board of any consumer goods company, and again, also in any food company.

We have briefly elucidated the traditional composition of executive boards of companies, with the usual suspects, such as MBAs, lawyers, finance experts, marketing experts, and the "alibi" technical representative. In the more recent past, the demand for having a greater female representation at board level can be heard loud and clear, and this rightly so. Now, in addition to all this, this book proposes, rather demands, that there is a strong and crucial need for having design represented as well. Maybe, just as a thought and to make the transition easier to digest, there is a bridge to be built between the female and the design element and these two new, increasing requirements could easily be combined.

It also has to be said that the design role at board level must not have an alibi character (like the technical or often still the female participation), but has to be a strong and equal partner, co-driving and co-deciding the direction and future of a food company, at eye level with the MBAs and marketing, and all the others. In turn, this could mean, actually does mean, that the executive board in its present form is a model of the past and more and more resembles the VW of the 1970s and not the most recent model: it is outdated and, what it always preaches to its employee base, namely change and flexibility, has to become a totally embraced mantra at board level as well. This is not a call for revolution, this just simply suggests to any company's management to go over the books and embrace the change that is happening wholeheartedly and on every level, including the top level.

There are not too many companies where designers are members of the executive board, maybe just a few in the automotive industry and the consumer electronics industry, but there is not a single example known to us—and to date—in the food industry. And this is exactly the point made here, in this book and in this particular chapter: establish strong, visionary, and decisive design thinking, in the person of an appropriate designer on the executive board of your company! From personal experience, this is obviously much easier said than done, in as much as it is easier to write it on paper, like here, than really do it, and execute the ideas that are proposed and are suggested as the most appropriate way forward. There is always a risk involved, namely that change will put at risk an organization that is apparently successfully operating and that does not require any change, at least not at the top level. At lower levels yes, but that's "for them to do."

So, how can we achieve such changes at the very top level? I do not have the silver bullet, and do not propose any such thing here. There is actually no such thing as a silver bullet, anywhere, and for anything. That's the sad reality, but there are many more arrows in the quiver that all of us can use to drive such change and drive it at any and *from* any level in the company. "Covert subversity," for instance, is one of the arrows that have been successfully used by the author. I would define "covert subversity" as the behavior that knowingly and willingly ignores hierarchical requests and demands when they are perceived as unreasonable, completely off the target, or outright irrelevant by an individual or a group. One may argue that top management never requests anything that would be unreasonable or off target or irrelevant because, in their position, they should know better and not come up with any of this. Yet, this is not the case and many years of personal experience have taught me that this happens time and again. And it's not just the wrong perception of the receiving end, it's often clearly the wrong message. And because it happens so often, people become insensitive to the request and respond with indifference or total lack of interest. This de-sensitization is similar to driving on a perfectly good road with speed limits that are far too low and make no sense at all: people believe that the authorities got it all wrong and yet, they stick to the limits (indifference), or they "challenge the wisdom of those up there" and respond by "covert subversity."

I do not suggest that people should break the traffic law, as this may quickly become a very expensive activity, however, I do suggest that in a corporate environment such covert subversity can make all the difference. Sometimes it is also called "skunk work," but then, names are just names and the important part is the behavior and the end result. If you can achieve surprising and striking results based on skunk work, you hold your head high and go and talk about it and advertise your or the group's success, and at the same time build the movement for change from the bottom up, to try to convince upper and top management that your success was not only based on your intellectual quality (or that of the group) and your perseverance, with a good portion of luck and perfect timing, but that additionally you have used great technology combined with even greater design and design thinking.

Such an approach, repeatedly applied, will ultimately result in change, on every level. It takes time, it takes guts, it takes patience, and it takes the proverbial "being at the right place at the right time with the right result" to convince people in your organization that change is also needed at the highest level and not just in your environment. Chapter 6 will go in much more detail with regards to the corporate reality in a changing business environment.

CONCLUSIONS

In this chapter we analyzed and discussed the role of design in the business world with special and purposeful emphasis on the operational and technical parts of the business. After all, the book is about the combination and interplay of design, technology, and innovation in the food industry. We especially described the relevance of design and enabling technology in the food industry, by discussing the various roles that design and technology play in the operational environment. We stressed and discussed the following areas:

- Design, the helper for business and technical, by discussing some past and present examples as to how design is perceived in a consumer goods company, with some emphasis on the food area.
- Design, the connector of business elements, gave us the opportunity to discuss and analyze the role of design in the value chain
- We led a more challenging discussion and analysis around the question of the many dimensions that design, as well as technology, can have and contribute to development and operations in the industry. We introduced the notion of the "n-dimensional design space" in analogy to the mathematical term.
- We discussed in some detail business interactions in today's consumer goods industry, human behavior, and how design and technology can help to improve these necessary interactions.
- The very important aspect of design's role in the supply chain was discussed in this section and we could show that clever design and smart technologies combined can help to substantially improve the performance of products throughout the supply chain, especially the outbound supply chain.

- Design as a counterfeit fighter was discussed and analyzed in great detail, with examples, and the size of the opportunity was presented and discussed.
- In our section on "the way forward" we introduced the concept of "down to earth" (DTE) design and the need for simplicity when it comes to successfully driving design concepts through a complex organization such as a food company.
- Finally, we discussed the concept of "design is management" and the importance of change in a corporation, change on every level that is, especially the need for the introduction of professional design thinking and a designer in person on the executive board of a consumer goods company.

TOPICS FOR FURTHER DISCUSSION

- Find examples for the role of design in your company and discuss them critically.
- Specifically elaborate on the role of design and technical in your company's logistics activities, with special emphasis on the outbound supply chain.
- Analyze in detail the role of design and technology as an enabler in the supply chain in your company.
- How would the "n-dimensional design space" look in your company, in your environment?
  Can you define it and what results could it bring?
- How does counterfeiting affect your organization? Do you have strategies in place that go beyond legal and take appropriate design and smart technologies into account? How important do you see the value of "speedy execution"?
- Discuss and analyze the importance of DTE/simple design in your company/ environment.
  List some good examples, combined with their success in the marketplace.
- Discuss the potential role of designers at executive board level and how this would play out in your own organizations. Discuss possible ways towards such a solution, provided you agree to the proposal.

REFERENCES

CBS Miami (2012). Fake food trying to make its way into the US. *CBS Miami, June* **18**, 2012.
A, Ciancio (2012), "Fake in Italy" branches out into toothpaste, soaps reports. *Reuters, October* **22**, 2012.
LeMatin Dimanche, June 2, 2013, p.40.
R. Martin (2009). *The Design of Business/Why Design Thinking is the next Competitive Advantage*, Harvard Business Press, Cambridge, MA.

6 The corporate reality in a changing world

Duty largely consists of pretending that the trivial is critical.

John Fowles, *The Magus*, 1965

THE DECISION MAKERS IN OUR SOCIETY: A "NEW ORDER"

Chaos theory and the well-known and often quoted "butterfly effect" suggest that tiny things, tiny events, and unimportant changes can have important consequences nearby and, more importantly, far away.

In a similar tone, Margaret Mead apparently once uttered the quote: "Never doubt that a small group of thoughtful, committed citizens can change the world. Indeed, it's the only thing that ever has." (http://www.interculturalstudies.org/faq.html#quote). There are probably many more similar statements out there, but all point to one thing and one thing only: it does not require many to make changes, even changes of a substantial nature.

Many books have been written on the topic of changes in a global world, or the changes in society, in families, in our communities, and other groups or entities. The point I would like to make here is that, in today's society, only a small group of people, the so-called decision-makers, can have a tremendous influence on the fate, present and future, of any community, small or large. One can argue that this wasn't very different in historic times, when kings and other nobles were the only ones to introduce change; however, this would be too shortsighted a view. Yes, it is true that nobility in Western as well as Asian societies had a great deal of influence, often without any possibility of opposition by the people, who just had to do what was asked of them. The only way to escape this dilemma was to organize a revolution with a very uncertain outcome. Let me, however, drive the point to the "democratization" of the instigation of change, which is a phenomenon of our days and which has come about with the rise of the communication era. We are fully in it: everything is public and everyone can propose and suggest just about anything, any type of opinion, any type of "personal news," and all this in real time. There is no intention to condemn or praise this phenomenon, it's just a reality, and it's called social media.

Food Industry Design, Technology and Innovation, First Edition.
Helmut Traitler, Birgit Coleman and Karen Hofmann.
© 2015 John Wiley & Sons, Inc. Published 2015 by John Wiley & Sons, Inc.

Chapter 12 will have a strong emphasis on the role of social media and the resulting networking, so I shall not emphasize this much more, other than to say that, despite the strong belief of many that politicians and governments, NGOs, and opinion leaders in academia, arts, or sports govern our society, I suggest that this is only a surface observation, and the real decisions and changes in our society take place "under the surface," led by an individual or a small group, perfectly using and playing the social media platforms, and by this probably influencing a lot more the direction and at what pace our society is heading and developing. Whether we like it or not, this is happening and we could cite many examples for this trend. Let me just mention a few here, and, interestingly enough, they are all based on the informatics revolution that began in the early 70s, and which might even have their roots in the students' movements of the late sixties.

For one, the invention of the personal computer, especially the increasingly powerful laptop computer, together with easy access to information, was a milestone in this democratization of change.

When Tim Berners Lee and his colleagues at CERN in Geneva, Switzerland invented the world wide web in late 1989, we were living yet another decisive milestone in this process. No revolutions by regular people are needed, these events are ready-made, ready-to-use revolutions in themselves, and society has joyfully embraced them. Wireless networking, the e-mail, and ultimately social media such as Facebook, Twitter, or Linkedin are the next big revolutionary events that gave the ordinary person tools at hand with which he or she could become as powerful as a politician, if not even more so. When all this became available, every single person could be connected through hundreds of thousands of nodes with other people in no time, they could call for meetings at public places almost instantly and use this platform, rather the enabler, to achieve all kinds of results, good and bad. Facebook parties are increasingly popular and all of a sudden, there might be several thousand people united to have a big dance party and neither the politicians, the government, or the police can do much about it, other than deploring their inability and the cost of damage the following day, as happened after such an illegal party with approx. 10,000 (!) participants in Berne, Switzerland on May 25th 2013. (http://www.swissinfo.ch/eng/swiss_news/10,000_fight_for_the_right_to_party_in_Bern.html?cid=32824660.)

Facebook is not responsible for this, it is just used as the catalyst and the uniting tool to organize such events, or any other events that an individual or a small group intends. It just needs a few to get to this tipping point of public following, and the multiplication factor is enormous; the idea of the public party may have been generated by just a handful of people, and a week or so later, more than 10,000 show up! And there are no costs involved, neither for the "organizer," or rather the instigator, nor for the people who attend; it just happens. One cannot say the same for politicians or artists or sports persons. Yes, many more spectators may show up to such an event, but just look at the organizational costs and how much you may have to pay to attend! It's much easier and equally, if not more, efficient to use social media to organize.

And why does it just have to be big dance parties and drinking bouts? Can't we use the social media platforms for much more productive, creative, and innovative goals? Yes, we definitely can, and perhaps it is even happening as you read these lines, yet, because it's less spectacular, maybe, and most of the time doesn't hit any newspaper headlines it's much less visible and noticeable. I truly believe that this will change and that our society will use such social media platforms, and the connectivity that can be achieved by them, in an increasing way in the future. This will, most definitely, have a strong influence not only on our society at large, but also on how corporations react to this trend and how they will embrace it and use it for their constructive purposes and goals.

THE DECISION MAKERS AND TAKERS IN THE CORPORATE ENVIRONMENT

As was introduced in the section above, the simple fact that our society has undergone quite some substantial technological revolutions over the last 40 years or so means that it can easily be predicted that corporations will embrace many elements of this technical revolution. As a matter of fact, the purely technical elements have been fully embraced for quite some time already, and it would be unthinkable today to work in other ways than the ones we are used to now, namely using all the technological progress in the business world. However, the other element, the psychological, the human element that stems from the easy availability of all these technologies has not been embraced by the corporate world to anywhere near the degree that society at large has already done. On the contrary, corporations still fear that what happens in society at large, namely the "democratization of change," could happen in their well-protected environment. Without wanting to hurt anyone's religious feelings, the situation of corporate structures, especially at the top level, could be compared to the Catholic Church's structure, namely, in order to keep holding on to the existing and historically grown power set-up, there is as little change as possible—everything is as it always was.

There is an old saying that goes like this: "If you do what you always did, you will get what you always got," and that is pretty much the situation at the top level of many corporations. The entire company may change, but the highest level is still structured as it always used to be. I cannot, of course, speak for all food corporations, but from personal experience, having met with many representatives of food and food-related corporations, I can say that this phenomenon is still widely observed. I would even go one step further: the more the top level asks anyone else in the company to change, to be flexible, to embrace new structures and approaches, the less they are ready to make any structural changes in their own top-level of organization. It is a rather strange situation that we find here, but it's an interesting reality nevertheless.

Psychologists would probably call this by an interesting scientific, preferably Latin, name; I call it simply fear of change and fear of loss of power. Change is especially necessary in any organized entity, be it myself, in my personal environment, in my family, in my community, in my leisure time, in my political party, in my circle of friends, etc., and particularly in my work environment. At work, change, and I mean constant evolution and the desire to always try out new approaches and new techniques, adding know-how, and using this appropriately is a true must; without it we would just stagnate and fall back. I am not advocating blindly following every new fad and hype, however I strongly suggest that without continuous learning, trying, and applying, corporations will not be able to succeed on the long run. It's therefore even more surprising that the people at the top of the organization are very resistant to change and hold on to their roles as much as they can, roles that have historically grown since the times of the lone entrepreneur who created a company of a still humble and over-seeable size, to the times where lawyers accompany every step of the process and supervise every action because the costs of liability for a wrong step could be very hefty, if not to say deadly.

So, apart from the lawyers, who are these takers and makers, who are the usual suspects that you will inevitably find in any corporation? Let's look at this in more detail via the example of a food company.

We have briefly mentioned this in the introduction, but let us go into a bit more depth and shed some light on the situation as it presents itself today. The relative distributions of roles may vary from company to company, and are largely driven by the corporation's desire to either be a more marketing-driven company (for instance, P&G) or a more tangible content-driven company (for instance, Nestlé). Depending on this distinction, there might be slight differences in the importance and responsibilities of the various roles, but the background and the typology of the people is always, boringly, the same.

The top role, the CEO, could either be an MBA with specialization in marketing, or a finance expert with specialization in marketing, or it could be a person with a law background, with specialization in marketing, or a technical person, with a deep desire to be a specialist in marketing … I believe that you get the picture: everyone is a marketing expert, as, apparently, that's the only real qualification that counts.

I must admit, in order to make my point, I have over-simplified. All these people, from my personal observation, have to have operational exposure with a proven track record of having successfully managed a business or a series of businesses.

Depending on the structural complexity of a corporation and whether it's organized solely as businesses, or solely as business units, or as a hybrid of business units and businesses (like Nestlé), the size of the executive board can vary. It is, however, clear that the hybrid solution does add some extra complexity when it comes to the composition of the executive board, as all businesses, individually or combined, and all business units combined want and need to pre-represented. It even leads to the situation that not all members of the executive board are at equal hierarchical levels, some are more senior than others. This may be a necessary "evil" and per se is not counter-productive. The situation can become even more complex, when, in addition to business units and businesses, the entire company is organized by geographical regions, which have to have responsible heads, and who, quite naturally, claim their seats on the executive board of the corporation. In case you have stopped counting, we have now three interested groups that have to be brought under one heading and direction, the direction of the company at large. Nobody said this would be easy, but this almost looks like there was a deliberate attempt to make it more difficult. There is, of course, history and reason behind such complications and we will discuss this in more detail in the following section.

SOME HISTORIC LESSONS IN COMPLEXITY BUILDING

As I just have outlined, executive boards of corporations, especially food companies, have organically evolved and grown into their present state. This is not different from any organization that evolves in size, geography, and economic importance over time. Many food companies have a rather mature age, some of them between 100 and 150 years, and are still going strong. Some of them are actually doing very well, so basically there is nothing wrong with anything, least of all with the composition of the executive boards and how they have evolved over time. Or is there? I believe there is, but then it's very difficult to put your finger on the exact spot.

I have made the point time and again that every executive board of every company, especially a food company, needs to be refreshed by the addition of design thinking and in the person of a designer.

So, this would just be another evolutionary addition to a board that has grown into the kind of organization that it represents today and that is driven by two main factors:

- Do my best to improve the success of the corporation
- Do my best to establish my role on the board.

This may sound cynical, but pretty much describes, in a simple way, today's reality. By the way, this is not board-bashing, we can probably say the very same for many levels within an organization, may be even swapping the sequence of importance; so, nothing outlandish here.

When companies are created, founded, established, or whatever term you may want to use for the first step, things are still rather simple as far as organization and structure are concerned. You typically find the driven, inspired, and inspiring entrepreneur, who holds many roles and little by little, as the company grows, becomes so stressed out by fulfilling all the required tasks that he or she will inevitably have to enlist the help of more people, more members of the still small company, still struggling to find the balance between financing, manufacturing, and selling with profit, without having lethal overheads and a high financial burn rate that straightforwardly lead to disaster. Anyhow, things evolve and very quickly—again, for the types of companies that we use as examples here, consumer goods companies, especially food companies—there appears some structure that basically consists of people responsible for:

- The company overall (the CEO)
- The development side of the company (the CTO or CSO)
- Manufacturing and operations
- Finance
- Marketing and sales
- Administration and human resources, the latter hopefully playing the role of what its name suggests: looking after the human resources, the people who work in the company, all of them!

The original entrepreneurs, in almost all historical cases of consumer goods corporations, were inventors, driven by an idea and having worked on a technical solution. We can safely say that the background of the "original CEO" was almost always technical. So where did we go astray and automatically accept that the CEO of today's corporation has to be a marketer? It's almost outrageous to believe that this always has to be and there cannot be any challenge to such a situation. Historically, the situation evolved in such a way that the MBAs and marketing experts told the technical people, the original entrepreneurs of the company, and even the present entrepreneurs of the company, that they should be happy with their roles, but that the real decision would be taken by the "grown-ups," namely themselves. This was a slow process and the erosion happened gradually, but surely. Luckily there are still many very successful companies around that have a very strong mix of technical and consumer understanding, with a technically savvy person at the helm. However, to my knowledge—and I may be wrong on this—none of the large food companies have a CEO with a technical education. In the discussion at the end of this chapter, I encourage the reader to critically illuminate this situation and discuss possibilities as to how the strong marketing and MBA orientation could be improved, obviously provided that you think that improvement is needed.

When we look back in the more recent history of large food corporations, the observation can be made that probably for the last 30 years or so, the now traditional composition of the executive board of said corporations has not been modified or even slightly touched up. From my personal observations in the industry, I have not seen any change and that is probably the most surprising conclusion of all: at the top level, structures have not changed for more than a quarter century. This is even more astonishing when we take into account the amount of change that has happened in the other parts of the companies, basically on every level from right below the board to levels 7 to 10 steps below! How can this be? And how does this bear up in the light of the mini-revolution that this book proposes, namely to inject some fresh blood, some really fresh thinking, paired with the highest responsibility, that a designer could contribute at board level?

And all this, when large corporations were totally re-organized and re-structured by abolishing vertical integration, only performing core tasks, and outsourcing those that were not business critical. This lead to the institutionalization of business units, the "McKinsey-ization" of the businesses in the early 1990s, to the embracing of the informatics revolution starting from the mid-1990s and the more or less willing embrace of the effects of a more or almost total global economy, paired with the need to show one's environmental responsibility and respect for sustainable and ethical business principles, even to the point of opening up one's innovation Pandora's boxes and mixing them with innovation coming from the outside, open innovation, which started in the early 2000s.

These changes and introductions over the last 20 to 30 years of course made it necessary for the roles of the individual members of an executive board to evolve with the changes, and the very same or almost identical structure of the board took on new members or existing members took on additional responsibilities. So, yes, roles have evolved, structure followed the inevitable, but remained globally unchanged, with the traditional roles of CEO, CFO, CTO, maybe COO, CLO (legal), Chief HR, and a few more chiefs for strategic business units, regional heads, and business heads. It must be added, however, that top management, with the exception of the introduction of the business-unit structure and the abolition of vertical integration, never wholeheartedly embraced any of the events described above. It can provocatively be said that the higher the person's rank in the company, the lower the preparedness and the readiness to embrace anything new, be it informatics tools or open innovation. Even globalization seemed to be a nuisance in the beginning until it was demonstrated that there could be important cost savings in smartly playing the global supply chains.

All this nicely describes the almost old-fashioned desire to hold on to old structures when it comes to my own, direct environment, but asking all the others to undergo substantial mindset and structural changes. If I am the executive board, I am at the highest level, therefore I can give the orders to everybody else. It's that easy!

"But the shareholders?" you will ask. Isn't the executive board accountable to the company's shareholders? The answer is simply yes, there is accountability, but it's more about the financial results and less about the company structure and how it is going about its business. As long as the expected results are there, as long as the company has a good and ethical standing and follows its sustainability and environmental goals, all is ok and no change is needed. If one or more of the goals, of the key performance indicators, are missed, then the big search begins as to where improvements need to be made and changes have to be introduced. But, apart from maybe the exchange of a person, nothing "ever" is structurally changed at executive board level. A bit strange, isn't it?

THE PROFIT MARGIN RACE

"Profit margin isn't everything, it's the only thing," could be considered the over-riding principle for all corporations. It may be considered a tough, even one-sided statement, freely chosen after Vince Lombardi's famous quote. Yet, having followed the evolution of the consumer goods industry and especially the food industry over more than 30 years, rather closely, I can say this with authority: margin improvement, as opposed to "margin dilution" is the number one driver of most businesses, even those that pride themselves in acting in a very sustainable and responsible way, not only towards their customers, consumers, and shareholders, but also towards their employees and business partners.

You must have come across many different definitions and explanations of "profit margin," and they are all made to make the whole thing look more complicated, even more enigmatic, than it really is.

Net profit margin (NPM) is a function of net profit (NP):

$$NPM = NP/Revenue$$

where NP is defined as = Revenue−Cost.

There are more definitions related to profit margin that take different parameters into account, such as before tax or after tax, or EBIDT (earnings before interest, depreciation, and taxes) or EBITDA (earnings before interest, taxes, depreciation, and amortization), etc. My opinion is that they are all there to confuse, but in accounting terms, and depending on the various accounting laws of the country in which a company does its business, all have certainly the right, or rather the need to exist.

This is not a business book, and the reader may have realized by now, at least, that I am not a business wizard, juggling freely with all possible terms and terminologies. One thing is for sure, though: net profit, in its simplest definition, is the difference between the net income from sales and the total costs of products sold. Net income is not a static number, but evolves over the accounting period, typically the calendar year or any other period that the company closes books. Figure 6.1 illustrates this in a very simple way. One can see from this figure that a company only really starts earning as the accounting year proceeds, as at the outset of the year, product- and production-related costs still outnumber the earnings, and only at the break-even point is there a real turnaround: the earlier this is achieved, the better the outlook for higher net profits for the entire year.

Figure 6.1 attempts to show in a very simple and graphic way that when a company really starts making money, the mid-year break-even, as depicted in the figure, in reality, is slightly misleading, as companies normally report profits from the first quarter on, although typically rather lower ones, compared to the following quarters. The reasons can be manifold, one of them being the infamous "end-year loading," i.e. selling a lot to the trade towards the end of the year, even if it might not be sold to the end-consumers, just to "make up the numbers." In most companies this is, however, strictly forbidden.

Another reason often seen, at least in the food industry, is that after the end-of-year holiday season, consumers are slightly saturated and may be more frugal as far as food

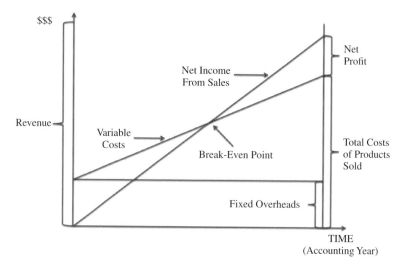

Figure 6.1   Net profit evolution during the accounting year (from P. Jacobs, personal communication, June 2013).

and beverage intake is concerned; maybe a direct or indirect result of New Year's resolutions? Figure 6.1 also very nicely illustrates that fact that most of "the numbers are made" towards the end of the year, or in other words, financial objectives are mostly attained in the last quarter of the year, although, quarterly projections of any balance sheets will always show profit that was made in real time. To complicate the issue, some food companies report other indicators, such as real internal growth (which is what it says, basic growth, without the addition of price increases) and organic growth (including price increases). This is typically the case for the Nestlé company and has been part of its business model ("The Nestlé Model") pretty much since the second half of the 1990s. The advantage of this type of indicator is that it is dynamic and not static, and shows everyone, including, very importantly, the employees of the company, the way forward and the targets to achieve in a very straightforward way.

The margin race, however, is not as clear and simple as it appears, and is very much culture dependent. It does not always go to the highest numbers, and this is for community reasons. Japan and the Japanese corporate world is such an example. The general opinion in Japan is to remain very conservative as far as margins and margin drive are concerned, and typical numbers are in the order of 5–7%, as opposed to Western markets, such as the USA or Europe, where the numbers would typically be more than double that figure. The major reason for this is found in Japanese society and its socio-political structure, especially the need to keeping employment numbers high, given the way Japanese live so close to each other, in densely populated cities, trying thereby to avoid social unrest in a rather fragile, yet highly balanced environment. This almost sounds like charity or at least very humanitarian, which in the business world is rather rare to find. However, calculating the social costs to Japanese society at large versus the reduced corporate gains, the net result apparently is in favor of lower net profits and better social glue.

This situation is almost unthinkable for the average Western or even South East Asian marketplace and financial analysts would react rather strongly. If a food company were to announce that, in order to "give back to the community," or any other reason for reducing profit margins, their shareholders would probably sell off, at best. It has to be said that Japan, due to its isolated and insular situation and the fact that many corporations in Japan are Japanese, and also due to the fact that a large portion of Japan's debt is local, such a low-margin strategy is much easier to execute than in most other parts of the world. However, it's an extremely interesting example of how businesses can also be run.

In the next section, I will discuss the topic of how corporations could be organized and run in very different ways, with new roles and "re-partitions" of tasks, to be simpler, more efficient, and ultimately even more successful. Some of the following might appear to be utopian; other parts are just a new way of looking at old structures.

VENTURE CAPITAL (VC): DECISION MAKERS BECOME RISK AVERTERS

There is a lot of money around in the world of investment. From institutional investors, venture funds, and private equity funds, to angel investors and private—public partnerships. And, quite understandably, every single one of these wants to maximize their investments and make money. However, there is nothing sure and safe about such investments. By definition, they are risky. In my personal experience, and always within the realms of the food industry, I was acting in three different roles within the investment world:

- Member of a technical committee in charge of judging the proper advancement of projects in which we (the company) had invested
- Technical expert for a group of angels investors, judging the technical feasibility of potential projects in which the angels wanted to invest
- Head of a food- and beverage-related start-up company in search of investment, from VCs, private equity, partner companies, angels, or whoever would please give us some money?

As you may guess from the latter, I was not very successful in securing money for my own venture, and this had a lot to do with the stage, the status of this venture. I was looking for investment in a company that did not produce and sell product yet. That apparently is a big turn-off for investors; they only want to invest in something that is rather risk-free and already produces and sells some kind of product at a certain value, the magical, often heard minimum sales value being of the order of $10 million. We could of course argue that if I am already selling for $10 million, I might not need the VC money anymore, which may or may not be true. What can be heard from people who are in such a situation, i.e. have already some ongoing business of the order of the above amount, they would prefer to keep the VCs out and try to do it on their own. This may be wise sometimes, but often money is dearly needed to ramp up production, purchase and pay for raw materials, or to invest more in marketing, so that it looks like it would still make a lot of sense for a VC to come in and spend the investment wisely.

From my role as member of a technical committee accompanying an investment into a project, I can only say that over a period of several years we did not really make any money, we didn't lose much either, and we were all excitedly waiting for the one great blockbuster that would make all our effort and hardship worth enduring. I can say, from personal experience, but also after having spoken to colleagues in the world of VC, that most of them, like me, are still waiting for the blockbuster that, unfortunately, never materialized.

The blockbuster may happen in some instances, but probably not too often. From personal experience, the industry standard regarding the evolution of the value of invested capital seems to be: 100 at the onset, 90–100 after three to five years of evolution. Not much loss, but overall no real gain either.

And, let me repeat, these are all investments into Phase B ventures, those that already have some ongoing business. None of the VCs or private equity investors that I have come across over a period of approaching 10 years now has ever shown any interest in investing in a Phase A venture, one that still is in the R&D phase of a potentially great idea. I should also repeat here that the experience is all based on food industry examples, which, in general, seem to be rather the step-child of the investors.

Here's a short quote from an article by A. Ball, analyzing the average return of VC investment.

He predicts a mathematical return of plus 25% over an observation period of 13 years from 1987 to 2000:

> *If an investment has an even chance of doubling or halving in value, it has a 25 percent mean return. For each dollar invested, you could make a dollar, or lose 50 cents. The larger the volatility, the greater this effect. More directly, VC investments derive their large average returns from a very small chance of a huge payoff.*
>
> (From: The National Bureau of Economic Research, July 2013; http://www.nber.org/digest/may01/w8066.html)

What he calls "large volatility" is based on the fact that when some of the invested companies go public, they typically make huge returns, of the order of 700 to 1,000%; but there are only very few of this kind.

Moreover, his observed time span is slightly longer than the one that covers my personal experience, during which time I have not seen anything like this, but, it has to be said again, my experience is food-industry specific.

My experience with the angel investors was more positive. It is no "Dragon's Den" experience; it's nothing of that kind, that's TV, almost soap. There may be angel groups that work that way, but without deep insight into the ramifications, technical hurdles, and possible legal liabilities, there was not one angel investor I have come across who would put out $100,000 in five minutes or so. However, if the angels agree to invest $100,000, for example, within a span of a few minutes, they always have this "yes, but" in the equation: let us double check technical and potentially legal hurdles before we finalize the investment.

It could well be that the "Dragon's Den" people do this as well, after the cameras are switched off, as it's just very detailed and sometimes hard legwork, and it does not make for good TV entertainment. The fact that angel groups spread their investment risk across smaller portions makes the process less hurtful to the individual investor. Moreover, what I saw in my angel group were investors of a certain kind, most of them wealthy, former business executives, or maybe wealth from family ("old money") who took real pleasure and interest in investing in something they saw as a great opportunity to help both sides: the asking party and not least themselves. I do not suggest that angel investors are philanthropists, certainly not, but there is a certain element of that involved in what they do. I have no absolute proof of this, but have understood from many discussions that the overall return on angel type investments is >100% after a similar period of time to that mentioned above for VC investments.

R. Wiltbank, in an article from 2012 wrote the following, which clearly supports my personal experience of positive returns on Angel investments:

> *The best estimate of overall angel investor returns from this data is 2.5 times their investment, though in any one investment the odds of a positive return are less than 50 percent. This is absolutely competitive with venture capital returns.*
>
> (From "Techcrunch.com": http://techcrunch.com/2012/10/13/ angel-investors-make-2-5x-returns-overall/)

Moreover, Wiltbank suggests in his analysis that the majority of these investments were in early stage ventures, which leads us to believe that early stage investments in general have better returns than the so-called safer investments in already active businesses, the latter just serving as multiplication, the former being real creation and generation of new businesses.

So, what can we learn from this? The obvious, namely that apparently safe bets are never safe and that shared risks involving many smaller investments from many individuals can probably lead to more clever investment strategies and better medium to long-term returns on such investments.

UTOPIA: FROM OLD REALITY TO A NEW REALITY?

Let's not forget the major topics of this book. In all the business mumbo jumbo we lost sight a bit of what this text has set out to achieve, namely describe, analyze, discuss, and reinforce the strong interconnectivity between design, technology, and innovation in the food industry. Later chapters will largely emphasize innovation and the pivotal role it plays in the consumer goods industry. However, in this section I want to emphasize the role of innovation and how we could have a fresh look at the industry's structures and organizations, as they are and as they could be, either transitionally or radically changed. The reader may find this strange, unusual, or even disturbing in some ways, but from long-standing personal experience, this is a necessary exercise in order to encourage a discussion amongst many players in the food industry and beyond. To have a better and more meaningful discussion, we should look, to some degree again, into the existing structures and ask the questions, why do they exist the way they are, and why do they operate the way they operate?

There are several answers to these questions; the more cynical answers are:

- We always did it that way
- We did not ask for your advice
- Don't bother us

Another, probably reasonable answer would be:

- It grew that way with the growth of the industry and has demonstrated its value.

Yet another answer could be:

- Because it apparently works, we must be doing something right, and we want to protect and even expand our own individual power zone within the organization.

You may want to find more detailed, more relevant answers within your own organization, provided there is no "Denkverbot" (not allowed to think) regarding these matters. More often than not, however, it is not Denkverbot within your organization, but rather no interest from the employees to discuss such matters or no perceived need to discuss any of this, to question the status quo and concentrate, at an individual level, on matters that just concern me and, perhaps, my immediate environment. Furthermore, to better understand the functioning of the industry, especially a food corporation, one has to better understand, to better read and decipher the important relationships between the major players, basically all players in such a corporation and its complex and sometimes almost chaotic interconnectivity.

Let us start doing this by first looking at the "value chain" of a corporation. The first and most important eye-opener is simply to recognize that there is no such thing as a value chain, as, by definition, a chain would have to be linear and there is certainly no linearity in the interconnectivity of the various value elements of any corporation. There may be a linear hierarchy in your organization, you even may have "line management," with straight and dotted connecting lines and career paths (simple or dual, or who knows what …), but there certainly is no chain, let alone a linear value chain. Undoubtedly, there are value elements, or functions or task groups or functional units or whatever they are called in your company, but they are linked together in a rather chaotic and non-linear fashion. Usually these are defined as complexity, perhaps sometimes giving you the impression that it's not really working, communication is not working, and this frustrates you, as the company could possibly be doing so much better if only … ? You must have had such moments of frustration, driven by the "if only" paradigm. Let us therefore introduce and later use another, more fitting, term for the non-linear value chain, namely the "value maze."

Let me say this very clear: this is no "call to arms," this is no incitement for revolution, but it is rather a call to use your own power of thought, followed by possible consequences that concern the content and structure of your organization. Two important words were used in the lines above: the first was "communication"; the second, "complexity." You must have heard time and again phrases such as: "We need to improve communication," or "We have to absolutely reduce complexity." Both of these phrases have a very true part to them, but are only half of the truth. The real phrases should be:

1. We need to start communicating, not just sending text messages or e-mails, or, almost as an excuse, have formal meetings, in which we should communicate.
2. We need to learn to manage complexity and use it to our competitive advantage.

These are the real issues, and neither one is appropriately covered and treated in today's corporate environments. My suspicion, which borders on almost total conviction, is that it's not treated properly because of the five answers that I elaborated earlier on and the strong hierarchical desire to not change anything that "works for me."

Regarding meetings and from long years of lived experience, I can say that there is almost a negative correlation between the number of meetings and the quality and intensity of communication: the more we meet, the less we really communicate; we download, we upload, we pretend to listen whilst we are looking at e-mail on our smart phones, chat with the neighbor about something unrelated, or correct a text that has urgency and priority. A minority of meetings actually do what they are supposed to do and what the name suggest: have people meet and exchange ideas and information in truly two-way communication.

Just a personal anecdote: when my second research director gave his introductory speech, more than 25 years ago, his first remarks were to say that "we have to improve communication." How can you improve something that you hardly ever do at all, in the true sense of the word? Whenever you hear this or something similar, and especially the phrase about "reducing complexity," you know that something's not right, something's fishy, and people do not want to talk about the real issues, namely what could and should be done to start a proper dialog, and what learning is necessary to master complexity in unique ways. There are other typical catch phrases or buzz words that people use when they don't know what to do to improve a critical business situation. One of these is: "back to basics." My understanding of this, based on long years of personal experience is: "I have no idea what to do next, so let's try the old stuff again, after all it worked in the past." Engineers, in their wonderfully dry and down-to-earth way have famous translation lists of what some engineering good-speak really means; I have not come across such translations in more business-related environments, and it might be the topic of a company-wide competition to find the real meaning of some of these catch phrases or buzz words.

I am putting some emphasis on this because it is good preparation, almost an exercise for the following discussion of the "Utopian new company structure." So, let us go back to the individual players in the game of the non-linear value chain, the "value maze." The real question is of course: what comes first in this value maze? The correct answer should always be, probably without exception: the consumer. At least, it's the politically correct answer and is certainly true. But there is not only the consumer who stands at the beginning of it all; there are at least two more players involved, namely "the trade" (or the customer) and professional operators. Other players could be consumer groups, NGOs, government rules, often differing from country to country ("the devil is in the detail"), lobbyists, the nutritional-medical professional, and, in the case of pet care, the veterinary profession.

The late Steve Jobs, undoubtedly a very successful entrepreneur, inventor, innovator, creator, and business leader, said many clever things and was mostly right in what he said and predicted. Here's a quote on the role of market research and consumers, which is totally contrary to how most or almost all peolpe in the food industry define the role of market research and better understanding of the consumer:

> We think the Mac will sell zillions, but we didn't build the Mac for anybody else. We built it for ourselves. We were the group of people who were going to judge whether it was great or not. We weren't going to go out and do market research. We just wanted to build the best thing we could build.

> When you're a carpenter making a beautiful chest of drawers, you're not going to use a piece of plywood on the back, even though it faces the wall and nobody will ever see it. You'll know it's there, so you're going to use a beautiful piece of wood on the back. For you to sleep well at night, the aesthetic, the quality, has to be carried all the way through.

> (Playboy, Feb. 1, 1985)

How about that? It's almost like marketing people in the food industry saying: I shall not go out and do market research and find out what the consumers like or dislike and how much they like it and make qualitative and quantitative predictions, because I, the marketing person (could, by the way, also be a technical person when working on a new product or packaging idea) know best, what the consumer wants; and, almost more importantly, I would not be a good marketing or technical person, if I did not have this deep conviction that I know exactly what "they" want. Sounds a bit arrogant, but it would be Steve Jobs' mantra translated into a food company environment.

Where's the truth, where's the middle way, you might ask. I believe that there is no middle way, there is only your way, i.e. you dare or you don't dare, and you are successful or you are not and subsequently ask for the proverbial forgiveness, not too many times though. I was personally involved in packaging projects with accompanying or rather underlying market research/consumer research as far as the new package was concerned. During the development phase we almost religiously responded to every consumer wish and new twist. When the research came back, we had "the best results ever, unheard of, a sure success." You may already have guessed what had happened next: the new packaging launch was a real flop and nobody seemed to know why that had happened. Until we applied our own common sense to the issue at hand and analyzed with the situation our own expertise and a bit of guts, and finally, rather cumbersomely, were able to correct this big mishap. Should we therefore throw over-board all market and consumer research? No, most likely not, but we should also not blindly follow what such research may suggest as it is always (actually ALWAYS) the result of a series of questions, more or less leading ones, that are asked by the researchers and often reflect some of our own misguided desires regarding the project.

In other words "we let think" and, to a large degree stopped thinking ourselves, handing our fates over to people who we basically control very little and pay very much. There is a way out, though, which I have had a few, quite substantial successes with in the past. Take your fate back, whichever level you work at and believe in the idea and the related development or marketing work that you do. Yes, do a few checks and balances based on common sense. Everyone (that is EVERYONE) is a consumer too and that gives you and your colleagues and your family and your friends the right to have an opinion. Most often, it is a very educated opinion because it's based on practical experience and stories of success and failure. If a package cannot be opened, why would you do a market research to find out how much this is a problem with the consumer? It is a problem and you can tackle it without having to ask anyone, especially not a market or consumer research company.

If you want to be entertained and want to see how consumers can or cannot open a package, you can do video ethnography, and watch and observe. But it's more of a candid camera situation, as you already know the answer and that the people are being tricked into a situation. It's a good laugh to see how they use their teeth or a knife or scissors or a letter opener or all of them, plus some brute force. It may also be a good laugh to see their frustration, but it basically shows you what you already knew: the package cannot be opened, "it's the package, stupid", do something about it and do it fast. No consumer research needed. There are many examples out there in the consumer goods industry of ill-functioning packages, badly tasting products and overly complex labels with illegible information, all of which can relatively easily be corrected without any consumer research. Yes, but what about new products, new ideas, something unheard of; shouldn't we ask the consumers? Well the quick answer to the latter is: if it's unheard of to date, then the consumers will not be able to tell you anyway, whether they want or need it or not. You are on your own, you have to decide, whether and how you want to move forward and spend money (not yours, but the company's) on such a product or packaging development. It's a tough call, but you will have to make it and stand behind it.

The reader hopefully remembers that we are still at the very beginning, and by the way, the very end of the value maze, namely the consumer, and yes, the professional operators and the customers of the food industry. Figure 6.2 depicts the value maze, indicating the complex interconnectivity of all individual elements by showing the major players inside and, to some degree, outside the company.

As the reader can see from Figure 6.2 above and the way it is depicted, the situation is rather complex, and sometimes looks as if the various elements of this value maze, all the players involved, are not really connected and do not really play together. This might not entirely be the case in many organizations, as they are all doing business, some of them even very successfully. So it looks like either at least some of the elements shown above are well connected and deliver as a kind of "state within the state" or that they have learned to appropriately manage this complexity and have adapted to it.

What has not been discussed yet is the varying importance of the individual elements, the various functions that we find in every enterprise. One could argue that they are the organs in a body, each of them has an important role to play. However, some of them are vital, e.g. the heart, and some of them have a lesser importance, for instance the appendix or the second kidney. Is R&D more important than marketing? Is design more important than engineering? Is sales more important than supply chain? I could continue asking these questions and the answers may not be as clear cut as in the human body: the whole would not work as well without the one or the other function as with it. However, that does not say anything yet as to whether any given function is of vital, core importance to the organization, or whether it could be outsourced and performed outside the corporation by competent

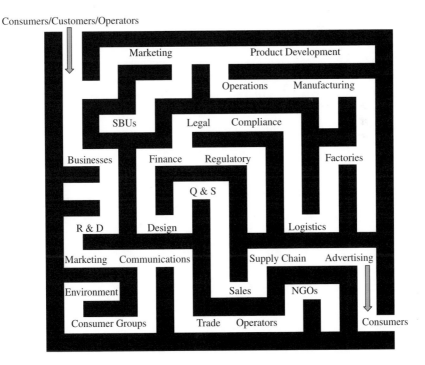

Figure 6.2 The "value maze": a picture with "issues."

organizations with a proven track record. One could imagine that legal functions, compliance, regulatory, or even logistics and supply chain, especially the transportation part of the last, could be outsourced and the company would perform equally well, maybe even better, at similar costs. As long as the company is its own manufacturer, and product and novelty developer, it appears that these functions are at the center of its interests and therefore should never be outsourced.

Alternatively, if the company decides to become a marketing and sales company, it may as well outsource R&D and manufacturing. We do not know whether such dissection will be the future of the industry and it will always depend on people, their ideas, and their strategic and tactical visions, as well as, almost most importantly, the history and heritage of any company. It is very difficult, almost impossible, to forget the latter and, from experience and long-standing observation, history may even be the strongest and most decisive driver when it comes to developing and executing strategic visions and directions for any given company, especially for food companies.

Let us try to think some utopian thoughts, center, left, and right of the old, the present, reality. It is well understood that the saying "don't fix it if it ain't broke," (Lance, 1977) has a very great importance in any situation in which one starts thinking of potentially changing a status quo, an existing situation, a working structure or organization. Yet, it is also clear that the best moments to have some radical thoughts about change and the future are at times of relative calm, stability, and success.

In the following, I have established four "what-if" scenarios in order to describe, analyze, and discuss various possibilities as to how a corporation could be organized and how its individual working elements could be interconnected or, maybe going even further, which of the elements would not be needed anymore and which new ones should actually be added. Although, the four scenarios still use all known elements and mix them in different ways, mainly by defining them in each of these scenarios, whether the function should or could be performed inside or outside the company, the final section of this chapter ("Topics for Further Discussion") will encourage you to potentially delete existing players or add new ones. Tables 6.1 to 6.4 describe and briefly characterize these four scenarios, defined by product-centric, design-centric, consumer- and trade-centric, and shareholder-centric companies.

Let me just add a few comments for clarity to each of these four possible types of company structures.

Table 6.1 The product-centric company

Functions inside the company	Functions outside the company
→ **Office of Strategy**: Decides on the major direction(s) of the company; includes R&D and product and packaging development.	Finances, Legal
→ **Office of Design**: Decides on product nature, its "soul," branding, and brand essence; is at the center of the product.	Advertising, Communication
→ **Office of Execution**: Has the task to "make" the product/packaging mix; includes Quality and Safety (Q&S)	Procurement, Logistics, and Supply Chain
→ **Office of Marketing and Sales**: Has the responsibility to market, move, and sell the product	Regulatory

Table 6.2 The design-centric company

Functions inside the company	Functions outside the company
→ **Office of Design Vision**: Decides on the major design and imagery of the company's products and packages	Finances
→ **Office of Design-Driven Product and Packaging Development**: Drives and manages R&D, packaging development.	Legal
→ **Office of Procurement**: Procuring raw and packaging materials is largely performed on a mix of design and cost drivers	Q&S, Compliance
→ **Office of Manufacturing**: Manufacturing driven by the initially decided design principles, also influencing best manufacturing practices	Regulatory
→ **Office of Marketing and Sales**: Design driven marketing and sales, also including communication and advertising	Logistics and Supply Chain; excl. Procurement

Table 6.3 The consumer- trade-, and operator-centric company

Functions inside the company	Functions outside the company
→ **Office of Marketing, Communications, and Advertising**: Decides on the major brand-related marketing, sales and advertising campaigns of the company's products and packages	Manufacturing and Operations
→ **Office of Design**: Decides on product nature, its "soul," branding, and brand essence.	R&D
→ **Office of Supply Chain, Q&S, Distribution and Sales**: Logistics is the major product quality and cost driver	Product and Packaging Development
→ **Office of Regulatory Affairs**: Compliance with country- and region-specific regulatory requirements	Procurement
→ **Office of Legal & Finances**: Driving the costs down and managing liability risks	

Table 6.4 The shareholder-centric company

Functions inside the company	Functions outside the company
→ **Office of Marketing & Sales**: Decides on the major brand related marketing & sales campaigns of the company's products and packages	R&D, Packaging Development
→ **Office of Procurement**: Procuring raw and packaging materials is largely performed on a mix of design and cost drivers	Manufacturing, Supply Chain
→ **Office of Communication & Advertisement**: Brand-relevant communication and efficient advertising become major strategic drivers	Q&S, Regulatory, Legal, Finances

First, regarding the product-centric company, I should add that this is probably my favorite structure: strategy, including R&D of product and packaging, design, including the full notion of brand essence, proper execution, including the all important element of Q&S, and finally marketing and sales are at the core of the company. Everything else is purely "mechanistic" and is needed to fully and professionally support the core elements. Therefore, I suggest outsourcing the elements of finance, legal, advertising, communication, procurement, logistics, supply chain, and regulatory. Sounds like a handful, and it is, but it's a radical way to seriously start thinking about what the company absolutely needs to have mastership in and what might be done more efficiently by outside resources.

Let's briefly look at the rationale behind the design-centric company. The explanations are the same: left side is core, right side is to be outsourced. It is very similar to the product-centric company; however, design becomes the driver of the organization and is at the heart of decision-making within the company. It is even more radical than the first variant and therefore has no big chance of being realized any time soon, if at all. However, it should be made the topic of serious discussions within your organization.

The third variant is the consumer-, trade- and operator-centric company. It has something going for it, actually quite a lot, as it would strengthen the links between manufacturer and customers; however, all development and R&D, most likely all front-end innovation work, would be outsourced in this variant.

Innovation would only be based on the feedback from the trade and operators. Although this is extremely important feedback, often over-looked in the present-day situation, it is far too mono-dimensional and not enough to ensure long-term sustainable and profitable growth for the company. Again, it should be the topic of discussion and analysis within your organization.

Lastly, let us briefly look at the shareholder-centric company variant. It's my least preferred, yet it has to be mentioned as there are always ongoing discussions, most likely in all food companies (and beyond), as to how better satisfy shareholders. Although most food companies frequently and credibly refer to a healthy balance between stakeholders and shareholders as an ideal situation, this variant needs analysis and discussion, and if only to be prepared with answers as to why this is not a viable company structure.

I let the reader decide which of these four variations, if any, resemble his or her own corporation the most. The assumption would be that none of these come close, mainly due to the fact that I have taken great liberty in outsourcing functions that I felt were not core in the individual scenarios. But then, we never know, and maybe you are already pretty close and rather advanced in your quest for a high-performance company. If that is the case, I would very much appreciate hearing about this from you!

CONCLUSIONS

In this chapter we discussed and analyzed the general and extremely relevant topic of the corporate reality in an ever-changing world. There was always change, at all times and in all places on our globe, however, it appears that there is more change than ever. Change creates fear and most of all uncertainty. In this framework of uncertainty we talked about the following sub-topics:

- The decision-makers in our societies and how such decision-makers, from a historic perspective, have come down from very few, high-ranking people (kings, emperors, presidents) to again, rather few, but no longer high-ranking persons, such as you and me, provided we use the capabilities that events such as the "industrial" and more recently the "electronic" revolutions have given us to hand. If it is cleverly used, few can move more and quicker than ancient nobility or even than democratically elected leaders. Whether we like it or condemn it, it has become a reality of our daily lives.

- We looked at the relatively unchanging roles of corporate decision-makers in an ever faster changing environment. They seem to take an ostrich approach: put my head in the sand and everything will be all right. We introduced the shear pressure of increasing complexity in organizations such as food companies as one of the reasons why business leaders can still find arguments for their own existence and the successful fencing off of changes to their individual roles.
- We discussed historic lessons in complexity and especially complexity building, and the reasons why the present leadership, and only they, can deal with complexity. It sounds more and more like an electoral campaign, with catastrophe scenarios and promises, than a reasonable and open dialog.
- We also discussed some very basic financial considerations in a very simplified yet understandable way. We talked about the margin race and the ongoing desire and struggle for higher profits as an important underlying factor as to why the present leadership of every corporation always finds excellent reasons why it is only they who can deliver. We also briefly mentioned the very different profit race in other marketplaces, with the example of Japan.
- In the next section we discussed the role of venture capital investments and capital investments in ventures of any kind, be it traditional VC or private equity, or angel investors. We discussed and analyzed the increasing risk aversion and suggested, based on checked facts, that higher risk and medium to long term leads to higher returns.
- In the last section of this chapter we attempted to elaborate on the topic of old business realities, the existing ones, to potentially new ones, not yet discussed, maybe not even thought of.

    We introduced the detailed complexity of the old value chain as being more of a value maze.

    This value maze served us as the basis to propose different utopian scenarios as to how a new corporation could be structured by also being very generous when it came to outsourcing apparently non-critical, non-core areas of the company.

TOPICS FOR FURTHER DISCUSSION

- Discuss and critically analyze the role of the new electronic media, the social media, with regards to decision-making in your direct local or regional environment. Note that Chapter 12 will extensively elaborate on social media, networking, and connectivity, especially in the corporate world.
- Critically look at the structure of your company, both in terms of human elements (the "people, who are our most important assets," as well as the leaders), but also how your company is organized: does it have all the elements it needs or too many/not enough? Discuss possible ways forward.
- Analyze and discuss "complexity" in your company and find your own answers as to how this complexity is managed both in your organization and by you. Is complexity

in your company a smoke screen behind which the leaders hide and justify the importance of their roles?

- How is profit margin explained in your organization? Is it talked about a lot, not at all, occasionally? How much do you know about it and how much, irrespective of your role, should you know and learn about it? This is an especially important question when you work in a technical function!

- How many ventures in which your company invested do you know of that actually made substantial profits? Have you ever been part of a venture capital investment meeting, be it at the onset or during the milestones-delivery phase? If never, maybe you should become interested? Do you have personal experience with risk aversion in your direct, project-driven environment? If yes, what can you, on a personal level, do to help alleviate risk aversion?

- Do you have an "old reality" in your company? Are there any ongoing or planned discussions you know of in your company that have the "new reality" as a topic? How do you see the "new reality"? Do you believe it would be necessary to at least move in this direction? Critically analyze and discuss the pros and cons of the four potential company scenarios at the end of the chapter. Is there anything "in it for you"? Do you see other scenarios? Please describe them and build your own "ideal corporation." Share the outcome with your colleagues and peers.

REFERENCE

T.B. Lance (1977). The Director of the Office of Management and Budget in Jimmy Carter's 1977 administration, *Newsletter of the US Chamber of Commerce*, Nation's Business, May 1977.

7 Design and technology: innovation is the connector

A rock pile ceases to be a rock pile the moment a single man contemplates it, bearing within him the image of a cathedral.

Antoine de Saint-Exupéry, "The Little Prince", 1943

DESIGN: BEYOND CONNECTING BUSINESS ELEMENTS

In one section of Chapter 5 we briefly talked about "design, the connector of business elements." Although I am convinced that design can and does play an enormously important role in connecting the various players in a corporation and may bring some order to the "value maze" discussed in the previous chapter, design plays an even more crucial role in influencing the technologies and the technological processes of any corporation, again, especially a food company.

When engineers develop or optimize new or existing technologies or production processes, they use the word design all the time. They even use the term "design freeze," and it comes quite naturally to them to apply design elements to their work. Design-freeze describes the moment at which no further changes are to be made to an agreed upon design solution, thus allowing those who have to execute it can do so on a solid and unchanged basis. However, it is a very special type of design that they use: the designing is especially related to drawing, drafting, and thereby optimizing a machine, a process, and combinations of equipment and manufacturing lines.

Although this looks like a very narrow view of design, it very clearly combines the already (many times) mentioned connection of "form and function," which engineers seem to understand almost naturally: without proper knowledge of this form-and-function paradigm and without meticulous application of this paradigm, the machine, the process, the factory line will just not work efficiently, and everything will cost more than it should, interruptions are more frequent, and line efficiency is down, and, and, … and management is very unhappy!

So, here seems to be a strong cost driver pushing for an optimal connection of technology and business, in which design plays a crucial, if not the most important role. It is a very neat example, which shows that design not only already plays an important role

Food Industry Design, Technology and Innovation, First Edition.
Helmut Traitler, Birgit Coleman and Karen Hofmann.
© 2015 John Wiley & Sons, Inc. Published 2015 by John Wiley & Sons, Inc.

in any type of corporation, but that this also applies to manufacturers that have their own manufacturing plants and pursue technological development of any kind, be it purely application driven in so-called "application groups" or a combination of such groups and a proper R&D organization, with a good mix of scientists and engineers working in these functions.

It is, however, interesting to observe that, although engineers clearly understand and embrace the notion of design as a critical part of their work, scientists do not feel and act the same way. Scientists (myself being one of that breed) have a different, if any, approach to design: "Design is for the others," "We scientists cannot do anything with design," or worse, "Design doesn't do anything for us." They are in a similar position to most other employees in a corporation, especially in a food company: "Design is for the others." And yet, there is much to be learned from the engineers, who have this quite natural and easy-going relationship with design.

Although most of the work that anyone performs in any type of company is linked to innovation, finding and applying new and efficient solutions to existing situations or creating new efficiencies in new environments, it is still not very well understood that innovation is really the glue that connects and unites technology and design, which are brothers (or sisters) "in arms," which go together like two identical twins and that one does not work, cannot exist, without the other. The topic of innovation will be at the center of discussion and analysis in later chapters of this book and it should be emphasized here, once again, that technology and design are linked together like Abbott and Costello, or Laurel and Hardy or Simon and Garfunkel. And I could go on trying to hammer in the importance of this dualism between technology and design, but I assume that the reader has by now not only understood the point I am making here, but also hopefully agrees to the crucial role of this connection.

In the following we will look at a few historical, as well as more recent, examples of this dualism of good design and engineering going hand in hand. The early steam engines (late eighteenth and early nineteenth century) are excellent examples of this connection and the closeness of engineering and design.

Figure 7.1 shows a historical Stuart steam engine No. 9, very nicely showing the attention to detail, the beauty, and the perfect functionality, including serviceability, of the device.

Figure 7.2 depicts very early details of a rotary engine after E. Cartwright in England, from a patent of 1797.

Figure 7.3 perfectly demonstrates the marriage of great engineering and probably even greater design, the almost mythical Riva Aquarama speedboat, created in the 1960s.

Figure 7.4 presents two more recent, food-industry-related examples of a production end-of-line buffer device from the Rotzinger company in Kaiseraugst, Switzerland. In chocolate production they are an extremely important device that helps alleviate line interruptions downstream, e.g. on packaging lines.

All these examples are based on a perfect combination of innovation, or rather innovative thinking and acting, with great engineering mastership and perfectly functioning design. This combination and the examples above cover a long time span from the late eighteenth century to the present time. They are so striking and ultimately so successful that there is no way around this paradigm of applying design and engineering principles, perfectly gluing together innovation and innovative thinking.

Figure 7.1 The Stuart steam engine No. 9; built by D. Haythornthwaite, England in approx. 1975. (Reproduced with kind permission of David Haythornthwaite).

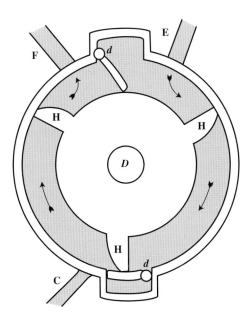

Figure 7.2 Rotary steam engine; E. Cartwright, England, 1797. (Reproduced with permission of Google).

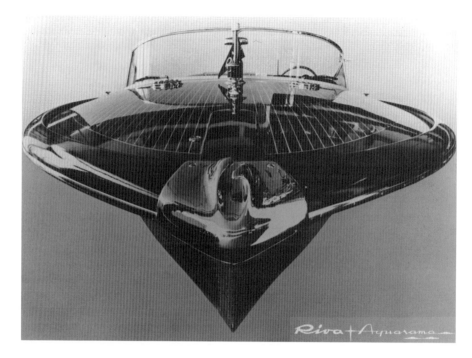

Figure 7.3 The Riva Aquarama Speedboat (From the Riva Historical Archive). (Reproduced with kind permission of the MSL Group, Milano, Italy).

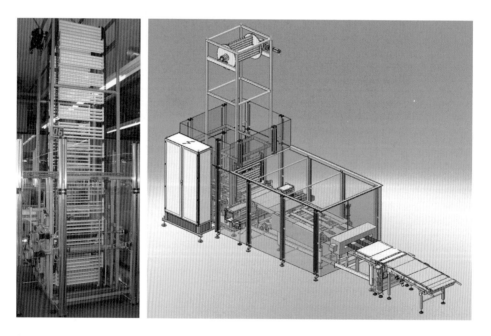

Figure 7.4 Rotzinger end-of-line buffer systems (accu-R-PFR 1 and 2). (Reproduced with kind permission of Rotzinger AG, Kaiseraugst, Switzerland).

HOW COMPANIES DEFINE THEIR BUSINESS STRATEGIES: A SHORT HISTORICAL PERSPECTIVE

Strategy has become the big buzz-word of the 1990, and still has the same clout, more than 20 years on.

Originally, this was a term used in the military, but is now largely used by business and in the corporate world. One short definition is the following: "A plan of action or policy designed to achieve a major or overall aim." (Google). The "plan of action" is the defining part. The "major or overall goal" is defined prior to this and at the onset, often as yet unrelated to the enabling strategy or strategies.

From many years of personal observation and lived experience I can confidently say that in companies that I am more familiar with, all of them actors in the food and beverage area, "strategy" is often confounded with "objectives" or "goals." Whilst "strategy" by definition describes a combination of set goals, why they are set, and, most importantly, ways to achieve them, the industry often just picks the goals part, more often than not also explains why it wants to go there—the strategic intent—but very often conveniently forgets to define how to get there. That's for others to define, or we define it on the go.

The important questions that we have to ask ourselves are: why does a company define strategies and subsequently attempts to pursue them accordingly? The questions seem to be trivial, but the answers are not as obvious as they appear. Let us look into the world of sports, especially team sports: plans are made on how to win the game, and each individual player has a defined set of actions and behaviors to follow religiously. They are called "the tactics of the game,", but they are not strategy, however. The strategy of a more or less successful soccer club, to use just one example, could be: Our intent is to win the Champion's League, within a time frame of three to five years, and in order to win it, we will: hire players X, Y, Z, in order to strengthen the defense, add one midfield player and one striker, all missing in the present team set-up. We are prepared to invest xyz millions in order to achieve this and our team manager will have the task of working out a system for the game that ideally combines the strengths as well as limiting the still existing weaknesses of the new team. The latter nicely leads over to the tactics of the team that ultimately are going to be applied so that the team will perform successfully and make its way up towards the top. Uncertainties, very much like in the business world, are the present and future environment of the sport, the other teams, potential future injuries of some players, surprisingly fewer spectators than expected, thus reducing the available income, losses in the early rounds of the championship and therefore not enough games being played to generate enough income to pay for all the investment that was made in the first place, etc. etc.

So, which of the various elements described above really is at the center of "strategy"? Most likely, all of the above, with the plan of how to achieve the goals at the center of all strategic consideration and planning. Companies, especially food companies, devise strategies and moreover translate these into communication snippets for the outside because a well-run business needs a strategy at its backbone, but also because the world of analysts expects to be told about the company's strategy. Again, here the same misunderstanding reigns: it's not the strategies that any communication to the outside talks about, but the final goals, which is totally legitimate and one of the better ways that any company performance can be measured.

Let us assume for a moment that everyone inside a food company understands what strategy really means and ultimately, at any level, becomes part of the appropriate execution, we then have to look into the origins of any strategy; in other words, who creates the company's strategy? Who or what is at the beginning of all this? Most of the leading food companies in existence today have a long-standing history behind them, some of them already being in their second century of life span. It can rightly be assumed that goals, milestones, strategies, and strategic intent one way or another have always been an essential part of such companies and how they ran and still run their businesses.

The really critical point is how to transform past and existing strategies in the future without upsetting the present organization, yet being ready to change just about everything if that ever became necessary. The solution, as always, lies in the middle ground: preserve the good parts and change or discard what's not working, and add whatever future survival and sustainable success will require. Who does that in large and complex organizations like a food company? The obvious answer is: the top management, it's their task to assure a successful present and future and that's what they are ultimately paid for. This is quite clearly also valid for any other type of consumer goods industry beyond the food industry.

From personal observation, I would say there are three major different, but ultimately converging, starting points for any strategy creation. These may not be the ones to be found in business and management textbooks; however, they are close to lived reality:

- Strategy created by the top leader of the company: one person at the helm has thought long and hard about it, proposes the strategy, and pushes it through the company.
- Strategy defined by a number of high-ranking company executives: a group of leaders proposes, discusses, and agrees upon the company's strategy and implements it by joining forces.
- Strategy as continuity: by default, the existing situation is confirmed as the valid strategy and is conveyed to every employee as the direction to go.

As said above, ultimately any of these different starting points may exist within the company and are used at different levels, depending on whether it's the strategy of the entire group or of a country or a region or a division. That is why convergence is so important and, at the end of the day, everybody talks the same strategic language with varying degrees of detail, or, to say it in business-speak, granularity.

FROM STRATEGY TO ACTION

The really important step is the one in which any type of strategy that was ultimately decided upon by management is put into action. It's relatively easy for a group of intelligent and relatively well-paid leaders to devise strategies, it's a totally different animal to ensure that the strategy is put into action and that every single member of a company understands, accepts, and transforms the strategy into what it really is: a pathway to success! Therefore, it's most important that strategies are not only written on pieces of paper and sent around to the employees to "take note," but that on every level of the company the whys and the hows and the whens are thoroughly explained, ensuring that all those who ultimately have to put all this into action become a convinced, enthusiastic, and highly motivated part of the realization.

The Nestlé example

One excellent example for what is described above is the fairly well-known Nestlé Group Strategy, initially published internally, more than 15 years ago (Brabeck-Letmathe, 1997). Beyond some very high-level goal settings and accompanying plans as to how such goals should be achieved by, for example, best product nature and availability, most appropriate and efficient consumer communication pathways, and R&D capabilities, which were accompanied by a quest for profitable and sustainable growth, as well as timelines of when these should be achieved, one other important document was created as a result of the Blueprint, namely the Nestlé Corporate Business Principles, a highly recommended reading material. All very down do earth, thought through, fitting in today's business environment, and, most importantly, totally actionable (Nestec Ltd., 2010). These principles are comprised of 10 major topics and are an excellent example of how a large food company complements its strategic intent with underlying and enabling business principles. Without wanting to go into much more detail, the 10 topics are the following, which speak for themselves:

1. *The Consumers*
   - Nutrition, health, and wellness (the main drive)
   - Quality assurance and product safety
   - Consumer communication
2. *Human rights and labor practices*
   - Human rights in business activities
3. *The people inside the company*
   - Leadership and personal responsibility
   - Safety and health at work
4. *Suppliers and customers*
   - Supplier and customer relations
   - Agriculture and rural development
5. *The environment*
   - Environmental sustainability
   - Water

From long years of personal experience, having lived through the development and establishment phases of these principles, it can comfortably be said that within the given company history and environment, these principles not only make a lot of sense, but strongly support the credibility and actionability of the strategic and tactical goals of the Nestlé corporation. It should also be said, especially as this chapter emphasizes innovation, the glue between design and technology that is innovation is written with a capital "I" in the Nestlé Company.

The Unilever example

When looking at other large, global food companies, such as Unilever, we find comparable, yet always slightly different strategies and plans of action. Unilever calls it "Our Compass Strategy" and this is broadly described under the definition of the "Unilever Sustainable

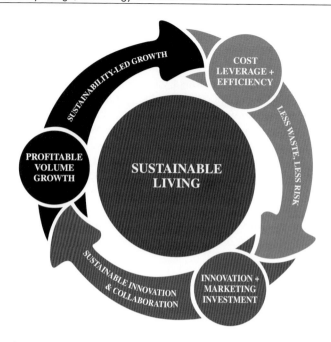

Figure 7.5   Unilever's virtuous circle of growth. (Reproduced with kind permission of Unilever PLC and group companies).

Living Plan" or USLP. The Unilever Compass Strategy is well depicted in Figure 7.5. It shows the important interactions between the various interconnected elements, as defined under their USLP initiative.

Unilever's USLP is composed of the following elements:

- Business strategy
- Brands
- Operations
- People
- Sustainable living
- Profitable volume growth
- Cost leverage + efficiency
- Innovation + marketing investment
- External recognition

Several differences can be seen between the Nestlé Blueprint for the Future and Unilever's Compass Strategy and their USLP. One important one is that whilst R&D was an important and equal pillar in the four pillars of the Nestlé Blueprint, the Unilever strategy mentions "innovation + marketing investment," under which one finds "investment in R&D." Whilst Nestlé mentions individual elements, such as human rights, labor practices, suppliers, and customers (the outside world), and specifically the environment, Unilever's plan talks about "external recognition," under which their "environmental and social performance" is mentioned. Yet other elements are very similar, almost identical, such as the need for sustainable and profitable growth, which, quite obviously are very universal drivers, or as the saying goes: "Without money, no music." (http://www.unilever.com/sustainable-living/ourapproach/ourcompassstrategy/index.aspx)

The PepsiCo example

Let us look at one other company, PepsiCo. In February 2012, PepsiCo announced "Strategic Investments to Drive Growth" with the following important strategic steps to be taken:

- Plans to increase advertising and marketing support behind its global brands by $500–$600 million in 2012, with particular focus on North America; going forward, it expects to maintain or increase that rate of support as a percentage of revenues.
- Multi-year productivity program expected to generate $1.5 billion of incremental cost savings by 2014 through optimization of operating practices and organization structure, including a reduction in force of about 8,700 employees, about 3% of global workforce.
- Company targets high-single-digit core constant currency EPS growth for 2013 and beyond after a transition year in 2012, in which it expects core constant currency EPS to decrease by 5%.
- Announces plan to increase returns to shareholders in the form of higher dividends and share re-purchases in 2012.

In the text following these four bullet points, there is much emphasis given to improved returns, productivity programs, investment—mainly marketing—in iconic brands, shareholder returns, and largely financially driven initiatives. Whilst there is nothing wrong whatsoever with any of these important elements, the word innovation or R&D did not appear a single time. However, "leveraging new technologies" was worth a mention and from the product successes that PepsiCo is having, one must assume that, despite being slightly under the radar for the outside world, PepsiCo is rather active in innovation and technology. It is also active in the world of design, having built its own design center led by Mauro Porcini, the new Chief Design Officer at Pepsi. He is building a world-class in-house design studio that leads and manages creative initiatives for all Pepsi products.

In conclusion, it can safely be said that, although communication emphasis and content is more financially and brand/marketing driven within Pepsico, inside the company much importance is given to the combination of technology and design, with innovation as the connector. (http://www.pepsico.com/PressRelease/PepsiCo-Announces-Strategic-Investments-to-Drive-Growth02092012.html)

The General Mills example

The General Mills (GM) Company is best known for its cereals business. They entered into a strategic partnership with the Nestlé Company and in the early 1990s created a joint venture under the name "Cereal Partners Worldwide" (CPW). The deal included the notion of clear geographies of distribution amongst the three corporations Nestlé, GM, and CPW. General Mills is one of the world's leading food companies, operating in more than 100 countries. With headquarters in Minneapolis, Minnesota, USA, in 2012 General Mills had fiscal net sales of US$16.7 billion. When looking at the corporate website of GM, it has to be noted that innovation is only mentioned once, and a lot of emphasis is given to shareholders, financial results, and environmental concerns and responsibilities, including such topics as waste reduction.

The following paragraph on supporting future growth strategies is to be found on GM's website. The main message is that GM pursues efficiency improvement plans resulting in cost savings that can be re-invested to accelerate innovation and growth.

General Mills announces actions to support future growth strategies
May 22, 2012

MINNEAPOLIS, Minn.—General Mills (NYSE: GIS) today announced a productivity and cost savings plan designed to improve organizational effectiveness and focus on key growth strategies. The plan includes organizational changes that strengthen business alignment, and actions to accelerate administrative efficiencies across the company. In connection with this initiative, the company expects to eliminate approximately 850 positions globally. Plans also include asset-related costs of approximately $13 million pre-tax associated with the write-down of selected production equipment. The company will record total restructuring charges of approximately $109 million pre-tax, reflecting one-time employee separation expenses and the asset-related costs. Approximately $94 million of these restructuring costs will be recorded in the fourth quarter of fiscal 2012, which ends on May 27, 2012. The remaining costs will be recorded in fiscal 2013.

Savings from these restructuring actions will be reinvested to support the company's future growth strategies and to accelerate innovation across General Mills' global business platforms.

General Mills continues to target fiscal 2012 adjusted diluted earnings per share of $2.53 to $2.55 per share. This guidance excludes the fourth-quarter restructuring charge, mark-to-market valuation effects, and *Yoplait* integration costs. (http://www.generalmills.com/en/ChannelG/NewsReleases/Library/2012/May/refuel_5_22.aspx)

The Kraft example

In September 2012, Kraft Foods laid out a new global growth strategy. Interestingly enough, the growth targets resembled very much the ones of the Nestlé Company, which is not really a surprise, as there is a certain reality in the food industry that hardly allows for very different growth targets when it comes to the upper limits. With annual revenues of approximately $48 billion, the company is the world's second largest food company. Although there is mention of savings in procurement, manufacturing and logistics, much emphasis is given to strong brands portfolio management, complemented by investments in marketing and innovation. It's a very similar combination of growth elements to PepsiCo's growth strategy plans. (Reliableplant.com, 2012; http://www.reliableplant.com/Read/26550/Kraft-Foods-growth-strategy)

I could add more examples here but the main message I want to emphasize appears to be rather clear: for most companies, innovation is not at the forefront of their business strategies, although innovation is an important, if not *the* most important driver for long term, sustainable, and profitable growth for any company in any type of industry, even more so for food companies. Moreover, technology or new technologies is rarely mentioned, if at all, in any major business strategy, and the notion of design is totally absent. This just shows how today's companies are run, especially by whom, and who is in charge: it's the financial interests. I am far from judging this to be a bad thing, and all this book intends to show is the need for a better balance between financial needs and interests, and the world of design, technology, and innovation in order to guarantee the best, sustainable future for consumer goods companies, especially food companies.

DESIGN AS AN INTEGRAL PART OF BUSINESS PLANS AND MARKETING STRATEGIES: A POSSIBLE REALITY?

When looking into business plans or even development plans in the food industry, the absence of design and design considerations is a recurring theme. There might be mention of design in some more detailed development or execution plans; however, in many years of personal observation, I have not seen any business plan that contained any substantial mention of design. In the company I worked for over many years, market (country)-relevant business plans were prepared, presented, discussed, and approved once a year. In none of these, design played any important role, if at all. One could argue that this is of no importance, this is how business works, we always did it this way, why change, why look into new ways of doing business, why anything at all … ?

I truly believe that all of these questions are probably never asked, because most business managers in the food industry do not even think in dimensions other than product success with the consumers, and what do I absolutely require, in order to achieve this success: good products, tasty products, healthy products, affordable products, efficiently manufactured products, cost advantage through procurement and best supply chain, great marketing strategists, social media, dedicated sales people, legal experts to fence off potential liabilities, well-calculating finance experts, and a general management that pulls all this together. This is of course a very top-level sketch of a successful food company, but, based on personal experience and observation, very well describes the present reality. Design is under the radar screen, technology is an enabler, and innovation or even commitment to innovation is often paid lip service because it might sound expensive to financial analysts.

Let us have a look at typical elements that a business plan should contain, especially when it comes to, for instance, a start-up company. However, many of the elements that are listed below would also figure in the annual business plan of any large food company.

The elements of a typical business plan

1. Executive summary: should be very short, ideally one page, maybe up to three pages, maximum. Should contain all pertinent information and should not only be concise but convincing.
2. A table of contents
3. The company summary such as:
   - For a new company or business: name, ownership, form of business
   - Mission, key successes, business and financial objectives, and stages
   - Location and facilities
   - The company's ecosystem (in which environment it operates) and possibilities for expansion

4. Products and services
   - The product(s)
   - Product features, including innovative technology, design, and advantage for consumers
   - Consumer interface (the package, a delivery system, etc.)
   - Development stages/phases
   - Competitive comparisons
   - The technology/technologies
   - Service and support
   - Future products and services
   - Fulfillment (what's in for the investor or the mother company once successful)

5. Summary of market analysis
   - Results of consumer research
   - Market size, growth, and trends
   - Target market segment strategy
   - Market needs: assessment and evaluation
   - Main competitors

6. Strategy and implementation summary
   - The "competitive edge"
   - Milestones and resources
   - Value proposition
   - Marketing strategy
   - Marketing programs and initiatives
   - Sales strategy
   - Sales programs and pricing strategy
   - Distribution strategy
   - As appropriate and required, possible strategic alliances with partners

7. Management summary on structure and resources
   - Organizational structure
   - Management and expert resources: description
   - Other permanent expert resources (as needed)
   - Management team gaps (if any) and how to fill them in the future
   - If required, description and role of advisory board

8. Financial plan
   - Important assumptions, such as consumer base, degree of consumer penetration, consumption frequency, price)
   - Projected sales
   - Status of maturity of industry segment, competitive environment
   - Vulnerability of economic factors and seasonality
   - Trademarks, copyrights, and potential legal issues, and how to solve them
   - Critical technological factors
   - Regulatory issues

9. Appendices (as needed), for instance:
   - Detailed results of consumer research
   - Comparative products in the marketplace and their retail prices
   - Projections on sales in as much detail as possible and credible
   - Worst case/best case scenarios; assumptions.

Although this may appear to be a lengthy list of almost 50 or so elements, it is important to realize that all these, and maybe some that in the reader's experience are still lacking, are crucial to successfully implementing a set strategy.

Design, technology and innovation play a critical role when it comes to products and services, and they should form important elements in every business or development plan. It clearly demonstrates that design as integral part of any business plan is not only a possible reality, but actually has to become an all natural, necessary reality, agreed upon by all involved in establishing such plans.

INNOVATION AS CONNECTOR OF TECHNOLOGY AND DESIGN

I have put this entire chapter under the heading of innovation as connector. Innovation does not only connect design and technology, as suggested here, but is without exaggeration at the origin of all activities in any corporation or even society at large. Without innovation, there would be no progress, there would be no improvements to existing situations, to existing technologies, to design in any area.

The slogan: "Innovate or die" really describes it well, although not every innovation is a blessing or necessarily good. Sometimes, the simplest ways, without too much sophistication through innovation can lead to important goals much more easily and quickly. When NASA worked on sending astronauts to the moon, it was apparently very important to develop all kinds of tools that the astronauts of the Apollo mission could use without fault, one of them a ball pen that would also work at zero gravity and was later praised as being usable by the average person who lies in bed and wants to hold a ball pen upside down; because the ink was under slight pressure in the pen, it would still flow upwards. Well, the Russians had a different approach, most certainly largely driven by lower budgets. They simply used a pencil. The pencil works under any conditions, gravity or not, downwards or upwards, and has the advantage of being much cheaper. By the way, there was a Russian motorcycle company "Ural," which once proudly sported the advertising slogan: "Unspoiled by progress." Not being a specialist in motorcycles, I cannot say how well this brand performed, but from insiders I have heard not too badly at all.

For those amongst the readers who had to do their doctoral thesis, like myself, under very restricted budget possibilities, it is quite an eye-opener to see how much restricted resources can and do drive innovative thinking and innovative solutions, ways around, to achieve the goal. As a student, one may even have learnt new skills like glass blowing when you had to build your own chemical equipment in the absence of money to buy ready-made glass parts. So, all in all, restriction is a great driver for innovation.

Figure 7.6 represents very simplified connections between the three elements of innovation, design, and technology, with innovation at the helm of the interactions and cross-influences.

Innovation inspires, enhances, refines, creates, and helps design to be meaningful and a strong element in the development and application of products and services. Innovation at the same time enables, allows, strengthens, steers, and improves technology in order to perfectly adapt it so that may serve products and services in a company.

Design and technology are connected through mutually influencing each other, strong co-operation, they lift each other up, they correct each other, and almost most importantly,

Figure 7.6   Innovation at the helm of connectivity between innovation, design, and technology.

they challenge each other for further uplifting and therefore jointly serve the corporation's products and services in quite unique ways. This is the true importance and meaning, the reason to be, of innovation: it has strong and substantial influence on anything else around it, and goes far beyond products and services, and can basically touch any part of the company, from supply chain to finances, from legal to procurement, from marketing to manufacturing, from sales to Q&S and would even, and probably especially, do Human Resources an awful lot of good.

INNOVATION IN DESIGN AND TECHNOLOGY CAN INFLUENCE HOW THE FOOD INDUSTRY OPERATES

The food industry, probably more than any other consumer goods industry, was and still is considerably conservative and, to put it nicely, rather slow in taking decisions, making changes, and implementing new, innovative products and processes. There are many very valid reasons for this situation, such as regulatory boundaries or health and safety considerations, and recognized food trends, which rather point to a "back-to-basics" situation like reducing ingredients or preferably using natural ingredients. Most of the new products that we can find in the food industry are often not based on innovation, but rather renovation of some existing products and packaging. There is fundamentally nothing wrong with this approach, as consumers expect their food and beverage products to be safe, tasty, affordable, and certainly based on common sense and well-established principles. "Don't try anything totally new, not yet well tested and well established on me," could be a good phrase that might represent the state of mind of the average food and beverage consumer.

It's not so much the industry that is conservative and risk adverse, it's the consumer that is asking for the well proven. The industry responds to this attitude and desire. On the other hand, everyone wants "innovation," I, the consumer, get bored with the never-changing product that I actually like, but still want to see in new and different flavor variations, expansions to other areas, such as from confectionery to ice cream or from powder to liquid, from a bigger package size to a smaller one etc. In principle, this is what smart renovation and good marketing is all about: present great variations to the themes of products and packages to the consumer and thereby not only keep my consumer base loyal to the product, but through these new variations on the theme additionally attract new groups of consumers and thereby increase my target market as well as the volume and size of the brand.

This is not to say that enhancing the quality of a product is always just renovation. There are quite a few examples in the industry that build on existing products and introduce very innovative features into the said product. Most examples where this happens are based on either very insightful consumer knowledge or over-riding strategic considerations of the company. When it became clearer and clearer that dairy proteins in infant formulae were the culprit for protein allergies being observed in small groups of babies that were fed them, it was an easy step to recognize that something should be done in order to correct that situation. In the case of Nestlé, the request to R&D was to come up with a feasible and affordable solution. In the late 1970s, Nestlé internal research had already worked on solutions and came up with the, in hindsight, obvious solution of offering partially hydrolyzed whey proteins in the formula, thereby reducing to a large degree the risk of dairy protein allergies in sensitive babies. This may sound trivial, but was by no means an insignificant development. Not only had the most efficient degree of hydrolysis to be found and tested, but technologies had to be developed in order to be able to manufacture such products on an industrial scale and, maybe most importantly, the taste of the product had to be acceptable not only to the baby, but also to the mother. It is a well-known fact that any baby formula and especially baby foods have to be "taste approved" by the mother.

Finally, the development was successful and the first hypoallergenic (HA) baby formulae was introduced into the marketplace in 1987 (Beba HA™). More recently however, a study conducted in Australia apparently shed some doubt on the claimed efficiency of the formula (Lowe *et al.*, 2011).

However, having had insight into consumer-related information and the evolution of the business and sales over the years, it must be said that the product definitely showed efficiency, simply based on the fact that mothers apparently were satisfied with the result and continued to purchase the product. A product that does not work cannot be continuously sold for over 25 years, even with the best marketing behind it. I leave the controversy for others to solve, yet the point here is that real innovation can be introduced into a product or group of products that have already been in the marketplace for a long period of time.

Real innovation in the form of actually really new products is rather rare in the food and beverages industry and the stories that can be told are rather limited in number. Here are a few examples.

Examples from the Nestlé Company

Nespresso®: having had some very limited exposure to the early development of the Nespresso® capsule in the mid 1980s, I have some personal insight into its history. The Nestlé Company had and still has as a very large and ongoing success story with its soluble coffee Nescafé®, by the way another truly innovative product invention, dating back 75 years in 2013. It can be expected that if anyone wanted to scratch at the base of such a big and successful product that the company management actually didn't really like it. Nespresso® in those days (until even the mid-1990s) was almost an Orwellian "thoughtcrime" and in the 1980s, the orginal Nespresso® team of four people had to almost be in hiding and were not supposed to make any waves. It was almost considered to be like toothache, it was bothersome, but would eventually go away, and yet, it didn't. Nespresso's success was not only based on the perseverance of the people involved; many more people eventually joined in and really worked on truthfully innovative concepts, from developing and designing the right combination of technologies, the capsule, its shape, its optimum material (ultimately aluminum was chosen) for best protection of the roast and ground coffee inside, how it would be filled and sealed in an industrially feasible way, how roast and ground coffee just coming out of a roaster had to be handled so that once filled and sealed the capsule wouldn't explode because of CO_2 still degassing from the grounds, and similar issues. This was not easy and it can be imagined that it took quite some time to solve all these problems and sell the first capsule, together with the dispensing machine. The machine and its development were yet another topic, and capsule and machine had to work perfectly together without failure. The first machines and capsules were sold in Switzerland in 1989, after having been tested in the "office coffee" sectors in Switzerland, Japan, and Italy since 1986.

The business model that was chosen was that of a club: the capsule could, and still can, only be purchased from Nespresso® directly, from the Nespresso® club, or its own beautifully designed shops, which can increasingly be found in the more fancy areas of large cities around the world. It is clearly a premium product, which is also reflected in the very careful selection of the different coffees that go into the capsules and its loyalty to origin, sustainability, and quality. Its rise to commercial success and glory resembled other innovative business ideas, such as Starbucks®: a very slow start and growth curve for the first few years, even as long as 10 years, and then, because the product delivers quality and lifestyle, a rapid rise to the "billionaires' club" of brands, i.e. brands that are worth more than $1 billion.

There are a few other, good examples out in the food and beverage industry of how design and technology influence the ways in which the industry operates. Interestingly enough, quite a few of these examples point to a similar pattern: a small group of dedicated and highly convinced people pursue an idea that either came from them in the first place, or, more likely, was given to them as a challenge by top management, responding to outspoken and, more often, unspoken consumer needs, but based on observation and anticipation.

Good examples are certainly the invention of Nescafé®, some 75 years ago, driven by a totally dedicated and convinced, and certainly stubborn, Max Morgenthaler in the pre-WWII Nestlé development laboratories in Switzerland. There was still a long and thorny path ahead to get to the technological perfection that soluble coffee presents today.

Another example, again from the Nestlé company and one in which I was personally involved, was the creation of the Nestlé Pure Life® water brand at the end of 1996 and its launch in Pakistan in 1998. I was asked by our then CEO a seemingly naïve question (although by definition, CEOs never ask naïve questions, only fitting ones ...): "Couldn't we make and distribute water like for instance Coca Cola would do it, by creating and protecting the concentrate and ship that to the end point for dilution and filling?" The short answer was: yes, we can do it. For water it would be the concentrated brine, the mix of the mineral salts that potable water normally contains. Such brine could then be shipped and the drinking water could then be reconstituted. The basic idea behind this, beyond shipping much less weight around the world, was also to propose a perfectly balanced and pleasantly tasting drinking water to consumers at very affordable prices, and especially in regions where safe and healthy potable water was rare and sometimes almost impossible to find.

We analyzed and proposed basic water purification technologies such as distillation and reverse osmosis, tested a multitude of different mineral salt mixes under the guidance of our nutrition expert colleagues and did hundreds of taste testing sessions with the help of our sensory evaluation colleagues. We were aware that natural water contains two things that reconstituted water would not contain: first, a mix of naturally present bacteria that hinder colonization of potentially harmful bacteria at least to some degree. This led us to do a very careful HACCP analysis, especially in the area of bottle filling. Second, such water is mostly deprived of silicate clusters, which allow for very high Ca concentrations, for instance, even at cold temperatures. Some natural mineral waters contain over 1 g of Ca per liter. We were aware that we could not achieve this and therefore had to be happy with a maximum of 200 mg per liter, actually even less than that, just to be sure to avoid any precipitation during cold storage. After more than 10 years of market presence of Pure Life water, it has become the biggest water brand in the world, found in numerous countries. The idea, or rather the question, was simple. Development was fast, and at first acceptance at the level of our traditional mineral water colleagues was non-existant; however, perseverance and luck turned this around, and the time from asking the question to launching in the first market was about two years, rather fast. Yes, admittedly, it was a simple development compared to some nutritional, functional products that may take years of development and testing, so we cannot compare easily, and shouldn't. Both types of products are necessary for a food and beverage company, as they ideally complement each other in a large and diverse portfolio of brands and products. In looking for other examples of what could be called "disruptive innovation," a category into which the above-described examples fit, in other food companies, it was more difficult to find comparative examples strictly in the food area.

Examples from P&G

One example that is often used, yet is outside the food area, is P&G's Swiffer®. Below is a quote from P&G's site, posted on October 5th, 2012:

Partnership with a Competitor Changed the Way the World Cleans

P&G engineers were working on a dusting tool to expand the Swiffer® mop line when C&D (Connect & Develop) teams came across a hand-held product being sold in Japan. They liked the sleekness and user-ease of the Japanese product—as did consumers in market testing.

The product's manufacturer, Unicharm, did not have the capacity, distribution or marketing might to take their product into other markets.

The resulting partnership enabled P&G to take the Japanese innovation global under P&G's established Swiffer brand. Now a market leader, Swiffer Dusters® are sold in 18 global markets.

C&D collaborations continue to expand the Swiffer® Family.

Other case studies are described on the site, 18 in total; however, all of them are outside the food area. After all, this is not surprising, as P&G has basically left the area of food, yet their approach to design and technology, largely described in other publications, was a basis for many of their product successes and led the way for food companies to follow in their footsteps. (http://www.pgconnectdevelop.com/home/stories/other-case-studies/20121005-swiffer-dusters.html)

Examples from the Unilever Company

Unilever still has a fairly large portfolio of food products and I searched, rather in vain, to find a more recent, food-related example of groundbreaking innovation. Unilever really led the way in science and technology when it came to developments in food lipids, especially in the area of lipid transformations and healthy lipids for margarine and spread products. This work dates back to the 1970s and 1980s, so roughly 40 years ago, yet still are valid to a large degree to this day. Examples of good design at Unilever can almost exclusively be found in relation to personal care products, especially in designing dispensing containers for cosmetic and body care products in a very functional and almost scientific way. http://www.altairproductdesign.com/(S(4xqt1lsqkq1m4gcjtydhpao0))/ResLibDownload.aspx?file_id=1087&&resource_id=2258&AspxAutoDetectCookieSupport=1

HOW COMMITMENT TO INNOVATION CAN INFLUENCE THE CORPORATE ENVIRONMENT: A FIRST GLIMPSE

It has been said time and again and has almost become today's mantra: "Innovate or die." And yet, there is still a great distance to cover from simply saying it, the lip service, to really applying it, living it, and making it part of the company's genes. Commitment to innovation has to be lived consistently and continuously: every day, day in, day out, always! It's this permanent focus on innovation that ultimately will determine whether a company is successful, long term, or not. This is especially true for food and beverage companies, and many companies have not only realized this, but do apply it, mostly on the surface, however.

I believe from personal observation and experience that the company I have worked for has always fostered innovation and an innovative climate, at least on a corporate message level. It is perceived by the outside world as an innovative company and was selected twice in a row by *Business Week* to be among the 50 most innovative companies in the world in 2009 and 2010. This was no small feat in two ways: first of all to become recognized in the corporate world not only as a company good to work for, but as a company that is recognized for its innovation capacity; second, and maybe almost more important, living up to the expectations that such a mention brings with it, being continuously innovative in a difficult environment, from a competition, as well as a financial point of view.

So, how can commitment to innovation influence a company, influence the corporate environment in meaningful and sustainable ways? I suggest that this very much depends on how innovation is defined at the level of the various functions in a company. The word innovation can mean many different things to different people, very much like "research," for instance. For a marketing person, research means market or consumer research, for a supply chain person it may mean finding new ways of cooling products in the warehouse or delivery trucks, for the financial experts it may mean looking into even more efficient ways to present relevant content of corporate reports to analysts, the engineer finds better ways of reducing line stoppages or increasing mean time between failures, and, obviously, the scientists believe that research is entirely his or her forte. The same is true for innovation: everyone sees a different picture, or maybe, sees the same picture, but with very different emphasis, content, ways to achieve, and even the ultimate result of innovation. Whilst general management sees their role as supporting innovation, or even demanding it from everyone in the company, everyone in the company sees this in a different light and yet, for general management, the message was clear: "Innovate or die!"

So, innovation, or rather the commitment to innovation, can strongly, but sometimes more strangely, affect the corporate environment in ways that may even resemble the proverbial building of Babylon's Tower. All involved seem to have the same goal, but they have a different conception, not only of the ways of getting there, but even to what the "where" actually is. There is lots of work to be done in order to unify the message and have everyone in the corporation understand the goals loud and clear and be given the freedom as to how they just might achieve the clear goal or goals. Again, from personal experience, I can safely say that the quest for innovation, or the message in which this is clad, is often so complex that there inevitably is room for misinterpretation. Later chapters in this book will extensively cover innovation, commitment to innovation, ways to innovate, and in particular ways in which innovation, real innovation, can ultimately be translated and executed into strong product and service successes in such complex environments as, for instance, the food and beverage industry.

CONCLUSIONS

In Chapter 7 we discussed, analyzed, and most importantly introduced innovation in general, and also exemplified in a few concrete examples. "Innovate or die!" is the old as well as the new mantra, and we discussed the following sub-topics in more detail:

- Design as a connector beyond the business element previously described in Chapter 5 towards connecting quite naturally with technologies, especially enhancing and optimizing engineering feats, such as machines and processes. We showed several historic examples outside the food and beverage industry, but also seemingly trivial ones inside the food industry.
- One important aspect discussed in this chapter was the importance of business strategies in corporations and the often-found confusion of strategy with objectives. Whilst objectives are of utmost importance, they only describe the end goal and what it is that a company would like to achieve, but fails to say how. This is where strategy transforms into a plan of action that, very similar to a good chess game, describes individual actions, necessary resources, and time lines, in order to achieve the set objectives.
- We discussed and analyzed the crucial part, "from strategy to action," realizing that often such strategies run the danger of remaining paper exercises. Several examples from companies such as Nestlé, Unilever, PepsiCo, and Kraft were presented and discussed, with the underlying question as to how prominently innovation, technology, or design figure in such business strategy papers. There seems to be a cultural difference in as much as European companies seem to put more emphasis on these elements. There is, however, no indication from the overall company results that this

has a big influence. Ultimately, what is done and executed is more important than what is written on paper.

- We presented and discussed the idea of making design an integral part of business plans and marketing strategies and looked in some detail into generic, but important, elements of a business plan, and where and at what stage design could best fit. Design, technology, and innovation must and increasingly will have to be crucial elements in business, marketing, and development plans.
- We introduced the idea, which is rather an existing reality, of innovation being the connector for technology and design, repeating the mantra, "Innovate or die!" We again discussed the need for simplicity in design (the "down-to-earth" type of design already mentioned in Chapter 5) as the crucial success factor, allowing design to become a full partner in innovation and technological development, and even beyond. We showed, in a simple figure, the possible interactions between innovation, design, and technology, and how they positively influence each other.
- We discussed and analyzed how innovation in design and technology may influence the ways in which food companies operate and presented a short list of examples, especially those that represent "innovation from scratch," as compared to renovation.
- We finally briefly introduced a first glimpse of how innovation and especially commitment to innovation may influence the corporate environment. We discussed diversity when it comes to the defining and understanding of innovation by different groups in the very same company.

TOPICS FOR FURTHER DISCUSSION

- Discuss how design influences engineering and engineers in your company. Discuss and analyze examples from your environment; compare these with examples from other industries.
- Discover and analyze your company's business strategy and understand it in as much detail as possible. Discuss possible differences between goals and action plans.
- Find and discuss examples illustrating "From strategy to action" in your company.
- Gain insight into your company's business or marketing plans. How much innovation, design, and technology did you discover in these plans? Discuss the outcome and propose ways in which you would improve the situation.
- Discuss the connector elements in Figure 7.1; add, delete, or modify, based on your experience.
- Discuss your personal experience and insight into the topic of how innovation, design, and technology have shaped your company and what steps you would propose to become even better.
- Find and discuss "innovation from scratch" examples in your company and compare them to examples given in this chapter or outside the food industry.
- How serious is your company's commitment to innovation? Do you experience confusion within your organization when it comes to defining innovation at different levels?
- Do you agree with the mantra "Innovate or die," and how do you see this in action in your company?

REFERENCES

P. Brabeck-Letmathe (1997). *Blueprint for the Future*, Vevey, Switzerland.

A. J. Lowe, C.S. Hosking, C.M. Bennett, K.J. Allen, C. Axelrad, J.B. Carlin, M.J. Abramson, S.C. Dharmage, D.J. Hill (2011). Effect of a partially hydrolyzed whey infant formula at weaning on risk of allergic disease in high-risk children: A randomized controlled trial. *The Journal of Allergy and Clinical Immunology*, **128** (2), 360–365.

Nestec Ltd. (2010). *The Nestlé Corporate Business Principles*, Vevey, Switzerland; http://www.nestle.com/asset-library/documents/library/documents/corporate_governance/corporate-business-principles-en.pdf

Part 2

Innovation: the much talked about, yet not well understood element

8 Innovation understood

We keep moving forward, opening new doors, and doing new things, because we are curious and curiosity keeps leading us down new paths. We're always exploring and experimenting; we call it imagineering—the blending of creative imagination and technical know-how.

Walt. E. Disney

INNOVATION AND CREATIVITY: THE FOUR STAGES OF VALUE CREATION

When asking people whether they are innovative, most of them will answer with a straightforward "yes."

When asking them whether they have creative minds, the answer will be more hesitant and not as clear-cut. People instinctively feel that they are innovative, even innovators, but they seem to have second thoughts about their degree of creativity. So let us look a bit into this connection between creativity and innovativeness and how these two influence each other.

Before I do this, let me add one very important statement that may not have been carved out clearly enough yet: innovation does not only happen in technical environments; it goes far beyond, and especially in the food industry covers all important domains of the business, including such areas as supply chain, marketing or sales. And design can and should equally importantly act as a connector between innovations in these areas and the underlying technologies, or techniques that help performance at a top level at any given moment.

Jeremy Clarkson in one of the series 10 episodes of the legendary "Top Gear" TV series came up with the following statement about his expert test driver "The Stig": "All he knows are two facts about ducks and both of them are wrong" (BBC, 2007). At first sight (or rather at first listen) it is just funny, but when you look at it more closely, it describes pretty well a good number of experts, people who should know a lot about many things, yet often only pretend. This is not only true for experts, but can equally be applied to the actions and behaviors of innovators and creators, in which case, this statement takes on a different meaning.

Food Industry Design, Technology and Innovation, First Edition.
Helmut Traitler, Birgit Coleman and Karen Hofmann.
© 2015 John Wiley & Sons, Inc. Published 2015 by John Wiley & Sons, Inc.

Steve Jobs once said the following about creative people and their creativity:

Creativity is just connecting things. When you ask creative people how they did something, they feel a little guilty because they didn't really do it, they just saw something. It seemed obvious to them after a while. That's because they were able to connect experiences they've had and synthesize new things.

The really interesting part of this quote is the part about the apparent innocence in this process: "I didn't realize that I was creative." In the same context, the fact that the "Stig," who only knows two facts and both are actually wrong, is creative and efficient in what he does, speaks to the same theme: innovators have a creative, almost playful innocence about them, they often don't even realize that they were being creative and innovative and found a solution to a problem that other people have long sought, and, in all their stressful and serious effort, were not able to see it. They just couldn't see the forest for all the trees.

This almost points to a situation that can be described as: the harder you purposefully try to be creative, the less it actually works. I have no intention in this part of the book and especially in this chapter to compete with the hundreds of books that have been written on how to stimulate creativity, what it takes to be creative, and how one could possibly learn to be creative and ultimately become an innovator. My proposition based on many years of practical experience and observation is that creativity and innovation cannot actually be taught, one cannot learn this, you either already have a mindset conditioned over many years that lets you be creative or else you have a tough time understanding and applying the process of creativity.

The same holds true for innovation. You cannot learn innovation, or in other words, if you are on the consulting end, you cannot really teach innovation, let alone creativity. However, what can be done is to stimulate people who are not so creative and innovative, and give them tools at hand that can help them to bring out whatever creative behavior they possess and apply this to the ultimate goal of creativity and innovation, which is creation and, ultimately the drive for execution. We will discuss "the innovators" in Chapter 9 and the most appropriate, and efficient tools that help innovators to become even more effective will be discussed in Chapter 10 in more detail.

Figure 8.1 shows the simple, yet extremely important, connection between creativity, innovation, creation, and execution. The most important realization for anyone, even an extremely creative person, is to accept that creativity alone is of no use, innovative thinking alone is of very little use, creation without a solid basis of creativity and innovation will not be successful, and that execution of creations that are not innovative is a very hollow and inefficient undertaking. As trivial as this all sounds, because these four elements in large corporations are most often disconnected and do not work hand in hand with each

Four Essential Stages that Lead to Value Creation

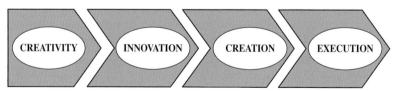

Figure 8.1 Stages of value creation.

other, the process does not work properly, or at least as efficiently as it could work if these steps became naturally interwoven in the organization of a company, especially in a food company.

It is not universally understood that these four stages cannot exist without each other and each step on its own is practically without value to any company. Although apparently understood by everyone in the organization, these stages, based on long-standing personal observation, are still very often, too often, handled in isolation. And this, despite better and convincing evidence that these stages have to work together and that there has to be a perfect handover and ultimate feedback at the end of each project, either a "post-mortem" or a "victory analysis."

The four stages are not necessarily identical with stages of maturity of ongoing projects within a company, but they rather represent stages of different states of mind, different mind sets of people working in such project teams. However, these stages are roughly reflected in the status of any project: ideation—testing—development—conceptualization—scale-up—pilot-plant scale—implementation—production. The terminology for these project steps may slightly differ from company to company. It is assumed that *creativity* contains the steps of ideation and preliminary testing of the concept, *innovation* covers development and parts of conceptualization, *creation* is all about scale-up and pilot trials and, finally, *execution* really stands for implementation and production or manufacturing in general. The fact that this doesn't happen all the time as it should in perfect harmony is very much explained by the reality of structures in large corporations. This is by no means an excuse; it's just stating a fact. Despite the wonderfully efficient electronic means of communication, very often silos or little kingdoms exist in large companies, geographically dispersed, regionally or globally operating, and the often needed proximity of members of project teams at various stages is no longer a given, especially with increasing financial pressure on travel budgets. I do not offer a magic bullet as to how this could be solved, but it is clear that there should be an appropriate and cost-efficient mix of project-related travel, as well as easy to use and simple web-based communication platforms.

The question might be raised with regards to the safety of confidentiality of any web-based communication platforms. It can be strongly suspected that even today's companies' internal intranet environment is not safe from foreign extrusion anymore. This book has no ambition to be a detective or spy novel. However, there should be serious discussion inside all companies, small or large, as to how data safety can be assured and communication platforms can safely be used. Here's a provocative assumption: increased travel budgets might be cheaper than loss of confidentiality and confidential data. But this is for others to calculate and discuss in detail, if at all. IT experts and specialists will obviously tell us that there is an IT solution for everything, but we should regard this with a little bit more caution, and, why not, good old-fashioned and down-to-earth common sense.

PEOPLE AND ATTITUDES

Who are the innovators? Is one born creative and innovative? Can I become more creative? Am I creative because I use my "right brain" more? Does the brain work differently in creativity mode? Can one be tested for creativity and capacity of innovation? Do creative and innovative people behave differently from less creative people? These might be some questions one could ask when it comes to creativity and innovation linked to people and attitudes. These might not even be the entirely right questions, or there could be more questions regarding this topic that are not posed here. I will attempt to go through these questions and try to answer them as we go along in this chapter, and go as deep and detailed as we possibly can to find answers, which will be more based on empirical and very personal observation and experience, having worked with a great number of creative and innovative people over a long period of time.

The first answer is: There are no simple answers to any of these questions, because no one has "I am creative" written on his or her forehead. We do not carry little cards with us, like blood group or organ donor cards, that say: On a scale from 1 to 10, my creativity degree is 6 or whatever. Maybe it would be a good idea, because it might render hiring of creative or purposefully uncreative people much easier, but then, who decides on this "level 6"? Moreover, it so happens that I might not be in any creative mood today ,but will be in my best shape tomorrow. From personal observation, preparedness for creativity and innovation almost follows a "circadian rhythm" of ups and downs, not necessarily during one day but over periods of several days or even weeks. People can be very creative for a period of time and, once they have been able to drive their creative idea, the subsequent innovation, the creation, and the execution through a complex organization such as a food and beverage company, then lose stamina, interest, or simply get tired and need a break from their peers' expectations that they can be super creative any time, any day, always.

Richard E. Cytowic, in his book *The Man Who Tasted Shapes* (Cytowic, 1993) discusses a brain condition called "synesthesia," a condition in which the human brain makes connections between seemingly unrelated senses. The most prominent examples are those of composers, who make a mental connection between a musical note, a tone, and a color. Cytowic in his book describes the rare case of a person that connects tastes and smells with shapes, or, in other words, a specific savory experience is connected, for instance to a square and edgy shape or a rugged surface or similar impressions. Apparently, the capacity to do so is, in a very basic way built into everyone's brain to differing degrees, but, because we do not need it and therefore do not use it, we lose this capacity of linking certain, seemingly unrelated senses and stimuli.

Moreover, Cytowic, in several chapters of his book, questions a number old views of "How the Brain Works" and in Chapter 19 proposes a "New View of How the Brain Works." The following is an original quote from Chapter 19:

> *The new view has five main points, which I will develop. In summary, they are as follows: (1) The flow of neural impulses is not linear, but parallel and multiplex, including transfer of information that does not even travel along nerves. Thus, it makes no sense to speak of hierarchy. (2) We no longer speak of localization as a one-to-one mapping, but of the distributed system: a many-to-many mapping in which a given chunk of brain tissue subserves many functions and also, conversely, by which a given function is not strictly localized but distributed over more than one spot. (3) While the cortex contains our model of reality and analyzes what exists outside ourselves, it is the limbic brain that determines the salience of that information. (4) Because of this, it is an emotional evaluation, not a reasoned one, that ultimately informs our behavior. (5) Likewise, all analogies of the mind to a machine are inadequate because it is emotion, much more than reason, that makes us human.*
>
> (Cytowic, 1993)

The importance here is not the exact ways in which Cytowic proposes that the brain works and the various parts of the brain are interconnected and work together, but more so the fact that there is no simple, linear way and the different parts of our brain control emotion as well as reason, not in hierarchical ways. This is an important statement, as it has a direct relationship to what shapes and makes creative and innovative people? Again, there is no easy answer and the very superficial answer should probably be: both, emotion and reason.

This view of how emotion and reason are interwoven, is really important to understand and absorb, as the basic fabric to better understand how creativity and capacity to innovate work and may be brought to the fore: by stimulating both parts, emotions and reason in balanced ways. They don't have to be stimulated in equal portions at all times, this is almost impossible; however, a healthy balance and even a healthier competition in stimulating them is the underlying secret to bringing out the best creativity and innovation in a person. This, in turn, means that being creative is not a monadic feat, it's not just coming from me, inside me and, surprise, surprise, here it is, I am creative now. No, creativity is always, and I repeat it, always, a reaction to a stimulus, which can come from inside the person or is an external stimulus, or mostly both.

Can anyone be creative and become an innovator? It's a difficult question and there may not be a good answer either. Before I go into a more detailed discussion of this question, let me quote my old high school teacher for geometry, who's name I have luckily forgotten, who used to say: "Geometry is a difficult subject: some have it, the others will never learn it." I was 16 years old then and, funny enough, he was pretty much right on the money. The "ones who had it" smoothly sailed through lectures and exams and the others "who would never learn it" really suffered and had really very bad times, they just hated it. One could argue that this is applicable to many other disciplines as well and it is probably true. Personally, I would not "have had it," for instance, for the law, but chemistry was easy (well, fairly easy) for me.

So, is the same true for creativity and innovation? Would my former teacher have said: "some have it and the others will never learn"? Maybe he would have, but I would like to discuss this in a more detailed manner. I will discuss and analyze some hiring patterns in the food industry that I observed and lived over many years and also the large variety of different characters that make the industry strong and have shaped it in the past and to this day. Given the size and complexity of this topic, I can only scratch the surface, but shall elaborate on recurring patterns for both the processes and the people. This will be specifically be dealt with in a later section of this chapter; however, it has a great importance in the context of creativity and innovation.

I still was dealing with the question, whether creativity and the art of innovativeness can be learned, whether it can be acquired through learning and lectures and if yes, how this might work. Chapter 10 will cover the topics of tools for innovation; however, these tools are of a generic nature and are equally applicable to everyone, innovator or not. The results of applying such tools may be different from person to person, but they should give everyone the chance to achieve important results. Again, this will be discussed in more detail in Chapter 10.

Consultants will tell us that everything can be taught, including the capacity to be an innovator. With all due respect, and from many years of observation and having worked with hundreds, if not more, individuals, the situation is rather like my high school teacher described it: the ones who didn't have it, never really acquired it. These people could, thanks to their own smartness and the help of their colleagues often get away with the fact that they were neither creative nor innovative, and could hide behind other important tasks such as project management or management in general. Don't get me wrong, both of these tasks are extremely important in any large (and not so large) organization, but they rather run under "red tape" than "moving us forward." Luckily, large organizations need such red tape activities, simply due to the fact that they are large. And, again luckily, this leaves room for the people who show strong traits of drive for creativity and innovation.

Let's not forget, as discussed earlier, that there are two more stages in value creation in a company: creation (the act of actually conceptualizing and realizing a concept) and execution (the act of actually making and manufacturing the concept in an industrial fashion). Although both of these stages could benefit from creativity and innovation, more often than not, they might also hinder, especially, proper and efficient execution. This is one of the reasons that R&D people who should typically be more creative and innovative, are often at odds with the more pragmatic employees that run the show in manufacturing. This is also one of the major reasons why handovers during the various stages of projects are often tedious, thorny, and filled with miscommunication, leading to rather sub-optimal results at the end of a project.

Great project management is quite naturally the solution to all this, and the right choice of great, unifying people, is paramount. From personal observation, I can say that very often the qualifications of project managers are based on traits such as: have made their marks in the past, are well spoken, seem to be liked by the boss (alternatively by the person who decides), seem to have answers to most open questions, have presence, and similar ones. What is often missing is the quality of being a unifying person and stepping back, letting other people shine, and encouraging them to stand up for their work, and, not least, teaching them how to "shine." Project managers do not necessarily have to be creative people and they themselves have to realize this, as good project management has a lot to do with pragmatism and efficient organization.

HOW TO BE AN INNOVATOR IN THE FOOD INDUSTRY

From the outside, the food and beverage industry is not really perceived to be innovative. This is also one of the reasons why no food company (strictly food and beverages) to my knowledge was ever amongst *Business Week*'s top 10 of the "50 Most Innovative Companies in the World". The Nestlé Company was, for the first time and two years running, in 2008 and 2009, amongst the top 50 of this ranking (38 and 36, respectively), but never any better. However, a study from "The Hay Group" reported in *The Nation* from June 2012 that food and beverage companies such as Coca Cola, McDonald's, the Wal-Mart Stores, Unilever, Nestlé, and PepsiCo were amongst the top 20 "Best Companies for Leadership," where efficient leadership was recognized as one of the most important factors for achieving creative, innovative environments and globally successful companies.

Innovation, Leadership Main Factors Behind Top Companies' Success: Study

The Nation June 4, 2012, 1:00 am

Samsung, Toyota, and Unilever top the rankings in Asia

Hay Group, a global management consulting firm, has released its seventh annual "Best Companies for Leadership Study", identifying the top 20 global and top 10 Asian companies for leadership and examining how they nurture talent and foster innovation in their ranks. This year,

General Electric topped the global list, followed by Procter & Gamble, IBM, Microsoft, and Coca-Cola. In Asia, Samsung Group, Toyota Motor, and Unilever led the ranking.

According to Hay Group's study, the Best Companies for Leadership (BCL) create workplace environments and processes that enable innovation to thrive. In fact, 90% of the Asia's top 10 companies reported that their leaders regularly celebrated innovation, compared with just 59% of other companies. In addition, 84% of Asia's top 10 companies reported that ideas from subsidiaries were just as likely to be implemented as those from headquarters, compared with only 63% of other companies.

"In Asia, there is a keen hunger to capitalize on the abundant business opportunities," said Mohinish Sinha, leadership and talent practice leader for Hay Group Asean, India, Pacific, and Africa and co-author of the BCL Asia report. "The Best Companies for Leadership recognize innovation is key to their future growth and sustainability in a fiercely competitive global market.

"Many companies prize innovation, but the Best Companies for Leadership approach it in a disciplined way by building agile organizations, promoting collaboration, and celebrating successes."

The global top 20 Best Companies for Leadership are GE, P&G, IBM, Microsoft, Coca-Cola, McDonald's, Accenture, Wal-Mart Stores, Johnson & Johnson, Unilever, Toyota, Nestlé, 3M, Southwest Airlines, ExxonMobil, PepsiCo, Siemens, Shell, Dow Chemical, and FedEx.

Asia's BCL top-10 are Samsung, Toyota, Unilever, Nestlé, Tata Group, IBM, Sony, P&G, Coca-Cola, and Petronas.

"There are plenty of ideas that we could learn from the best companies in driving a culture of innovation," said Thachasha Phuangpornsri, leadership and talent practice leader, Thailand. "Innovation does not happen by chance anymore; the highlight this year indicates that best companies focus on promoting an environment that encourages meaningful innovations throughout the organization. This is seen as a crucial way to survive and drive their market leadership."

Google's 80/20 Innovation Model is a great example who walks the talk in terms of "promoting an innovative environment and culture": Google's 80/20 "innovation time off" (ITO) policy encourages Google employees to spend 80% of their time on core projects, and roughly 20% (or one day per week) on "innovation" activities that peak their personal interests and passions. The questions that remain are, how we can translate this example which works well for developers, engineers, and other creative types to the rest of the organization in particular to management, factory, administrative assistants, etc.

Also, in Chief Design Officer of PepsiCo Mauro Porcini's talk on "how to inspire a culture of innovation" through "design thinking", a form of creative problem solving, he emphasizes on the people/consumer and to invest in understanding and connecting with them. "The emotional connection that people have to a product or brand should drive the design. Design thinkers, he says, aren't necessarily designers themselves. Instead, they are people who are curious and listen with humility. By the end of the day, customer loyalty comes from how a product or company makes a person feel and the story it tells them about themselves. Companies do not design products, they design experiences.

Innovation revolution

When it comes to innovation, Hay Group's study also found that Asia's top 10 were driven by future customer needs and not just serving the current ones. In fact, 90 per cent of them report that employees—not just managers and leaders—constantly discussed customers' future needs, compared with just 56 per cent of other Asian companies.

When faced with difficulties, Asia's top 10 are willing to postpone short-term profits to continue investing in innovation (71%) and treat negative feedback and performance difficulties as learning opportunities (77%), compared with 53% and 56%, respectively in other companies.

"The ability to translate future trends into new product offerings has kept Asia's best ahead of their competition," observed Vanessa Cen, leadership and talent practice leader in Hay Group Northeast Asia and co-author of the BCL Asia report. "The willingness to bite the bullet now by putting off short-term gains so as to continue innovation investment is their hallmark.

"In fact, Asia's top 10 (71%) do better in this aspect than even their global top 20 peers (65%). Coupled with their organizational resilience, this enables them to bounce back quickly from setbacks, and the Asia top 10 are a force to be reckoned with."

Global best practices

However, the co-leaders of the study warned that Asian companies should not rest on their laurels as they have much to learn from the global top 20 companies in the following areas:

Develop an organizational structure that enables quick communication. This is fundamental to organizational agility, decision-making frameworks and responsiveness to market changes.

(bolded by the author).

The Asian tradition of directive leadership will not work to foster innovation. Everyone must be expected to lead, even those who have no formal position of authority, and Asian bosses must learn to delegate their authority and decision-making power.

Create personally meaningful work. Having an innovation strategy is no guarantee of success, and the most innovative companies are interested in discretionary effort. Therefore, strategies must be decoded vertically and horizontally so that personal interests are aligned with corporate and interdepartmental goals.

Hay Group said it would hold the World's Most Admired Company (WMAC) 2012 Forum in Bangkok on July 4, featuring guest speakers from Banpu, Nestlé, Charoen Pokphand, and Toyota."

(The Nation Multimedia, 2012)

This view from the outside only partly reflects the reality, namely the simple fact that the food and beverage industry is extremely creative, although not in any spectacular ways, and therefore, it is not perceived as being amazingly innovative, like, for instance the computer, software, or automotive industries. Both computers and cars are seen as technological feats by the public; however, food has to be healthy, safe, affordable, and definitely not fancy; no chance of being perceived as innovative by the consumers. We have already discussed great examples of innovation in the food industry in Chapter 7. This present chapter and section is all about people, especially innovators and how they act and survive in this industry. Given the fact that innovation in food is not a prime expectation of consumers, innovators have not the high standing, compared to innovators in other industries, which makes their lives a little tougher and they often have to almost excuse their creativity and innovativeness in a rather down-to-earth and pragmatic context.

The well-documented story of Max Morgenthaler, the "lonesome inventor," who stubbornly pursued an idea, making a convenient, easy-to-use, stable, and good-tasting soluble coffee beverage is a great example of what it takes to be an innovator and especially an inventor in a large company, where many players potentially pull in different directions, often not even consciously, but because they think their direction, in the given environment, is the best. Dr. Morgenthaler, against all odds and against many in the company, after several years of trials and errors developed, in an approximately two-year time span, the famous product: Nescafé® instant coffee. As early as 1941, Henry Miller, in his book *The Colossus of Maroussi*, described that in his train travels with Lawrence Durrell, they drank Nescafé®! There are many more examples of successful innovators in many food companies and even more outside the area of food and beverages. The traits and personal characteristics of all such innovators are most often the same: stubborn, determined, obviously competent, open to looking across the fence (!), ready to embrace deviations and with a good nose for the lucky moment; and yes, being in the right place at the right time, working on the right subject.

INNOVATIONS AND INVENTIONS IN FOOD AND BEVERAGES: A SHORT HISTORICAL OVERVIEW

When you ask the question: "What is innovation" you may get many different answers and they may all point you into the right direction and describe well what innovation means. Before we delve in more detail into the history of innovation and inventions in food and beverages, let me add probably the simplest of all definitions: "Innovation? When you see it you recognize it." This is pretty much the recurring theme of all examples that are listed in the article below: "Oh yes, I can see that this was really innovative and ground-breaking!" In June 2012, the following article was published by the "Atlantic Media Group." It very nicely and briefly describes the 20 most important inventions in the history of food and beverages and is quoted here:

The 20 Most Significant Inventions in the History of Food and Drink

Megan Garber, *The Atlantic*, September 14th 2012

The Royal Society, the UK's national academy of science, had a question: What are the most meaningful innovations in humanity's culinary history? What mattered more to the development of civilization's cultivation of food: the oven? The fridge? The plough? The fork?

To answer that question, the Society convened a group of its Fellows—including, yup, a Nobel Prize Winner—and asked them to whittle down a list of 100 culinarily innovative tools down to 20. That list was then voted on by the Fellows and by a group of "experts in the food and drink industry," its tools ranked according to four criteria: accessibility, productivity, aesthetics, and health.

Below, via Edible Geography's Nicola Twilley, are the ranked results of that endeavor. These are—per the eminent body of the Royal Society—the top 20 innovations in food and drink, from the dawn of time to the present day.

1. **Refrigeration**. The use of ice to lower the temperature of and thus preserve food dates back to prehistoric times. Machine-based refrigeration, however, was developed as a process starting in the mid 18th century and moving into the 19th. Domestic mechanical refrigerators first became available in the early 20th century. Throughout its long history, refrigeration has allowed humans to preserve food and, with it, nutrition. It has also allowed for a key innovation in human civilization: cold beer.
2. **Pasteurization/sterilization**. Useful for the prevention of bacterial contamination in food, particularly milk.

3. **Canning.** Developed in the early 19th century, canning is a method of preserving food by processing and sealing it in an airtight container. Canning provides a typical shelf life ranging from one to five years. It also provides a fun weekend activity for humans who live in Brooklyn.

4. **The oven.** The earliest ovens, found in Central Europe, date from 29,000 BC, and were used, at times, to cook mammoth. Their more contemporary counterparts, gas ovens, were first developed in the early 19th century and were used, at times, to cook buns.

5. **Irrigation.** Irrigation is the artificial application of water to land or soil. It is used to assist in the growing of agricultural crops, and in the revegetation of disturbed soils in dry areas. This is particularly useful during periods of inadequate rainfall.

6. **Threshing machine/combine harvester.** Invented in the late 18th century, the thresher brought more industrialization to farming, allowing for the mechanized separation of grain from stalks from husks. The device was originally called the "thrasher." Prior to its invention, farmers had separated grain by hand, with flails.

7. **Baking.** This is a food cooking method that employs prolonged, dry heat to cook food. Baking acts by convection, rather than by thermal radiation, and is typically undertaken in ovens, in hot ashes, or on hot stones. Its results can also be facilitated by invention #14.

8. **Selective breeding/strains.** Selective breeding is the process of breeding plants and animals for particular traits. It allows humans to manipulate natural selection among the plants and animals they consume in order to produce food products that are genetically stable. It is also the reason that a bull named Badger-Bluff Fanny Freddie has sired thousands of the dairy cattle in the United States.

9. **Grinding/milling.** Grinding is the process of grinding grain or other materials in a mill. It produces, among other things, flour, which is the main ingredient of bread—a staple food for many cultures. The milling of grain has been a practice since 6000 BC, enacted by millstones and similar implements—and replicated pretty much the same way until the late 19th century brought the advent of the steam mill.

10. **The plough.** A plough is a tool (or, more commonly now, a machine) that cultivates soil in preparation for sowing seeds. It has existed, in some form, pretty much since the dawn of recorded history, and represents one of the major advances in agriculture. In that regard, it facilitated the rise of sedentary human civilization.

11. **Fermentation.** Beer. More formally, "the conversion of carbohydrates to alcohols and carbon dioxide or organic acids using yeasts, bacteria, or a combination thereof, under anaerobic conditions"—which leads to such products as alcohol, wine, vinegar, yogurt, bread, and cheese. Mostly, though: beer.

12. **The fishing net.** Fishing nets have been used since the stone age, with the oldest known version made from willow and dating back to 8300 BC. The nets are still in wide use today, and currently include casting, drifting, dragging, landing, trawling, and leg-wrapping varieties.

13. **Crop rotation.** The practice of growing a series of dissimilar types of crops in the same area in sequential seasons.

14. **The pot.** The food container was an invention of revolutionary simplicity—one that allowed for such revolutionarily simple acts as boiling water. Which led to innovations like tea and pasta and cook-in-a-bag omelets. Additionally, "the pot" is the culinary innovation with, by far, the most intriguing disambiguation page on Wikipedia.

15. **The knife.** Including, ostensibly, the Ginsu.

16. **Eating utensils.** Excluding, ostensibly, the spork, a spoon/fork hybrid.

17. **The cork.** The cork allowed for the production of wine and beer. Cork's elasticity and near-impermeability make it ideal as a material for bottle stoppers.

18. **The barrel**. Same deal: wine and beer. While barrels store other things, too—water, oil—they made this list, ostensibly, because of the booze. As a technology, barrels also led to the advent of one of the least-scary forms of clown yet invented.
19. **The microwave oven**. Often colloquially shortened to microwave, this is a kitchen appliance that heats food by a dielectric heating process in which radiation is used to heat the polarized molecules in food. The microwave is notable mostly as a gateway technology, leading to the culinary innovation of the late 20th century: Easy Mac.
20. **Frying**. The cooking of food in oil or another fat originated in ancient Egypt around 2500 BC. Civilization recently combined frying with all of the above 19 innovations to create the product that will probably, somehow, lead to its end: deep-fried beer.

(From: The Atlantic Monthly Group, http://www.theatlantic.com/
technology/archive/2012/09/the-20-most-significant-inventions-in-
the-history-of-food-and-drink/262410/)

The really interesting part of this publication is the fact that the different methods of food preservation and protection, led by refrigeration, made the top of the list. Agricultural tools and methods were the second most important aspects, followed by all other operations, utensils and tools that make food and beverages not only stable for a longer period of time, but make them tastier and ultimately, altogether healthier.

Other sources describe similar historic feats and inventions in the area of food and beverages, including agricultural advances. A very instructive site is "Enchanted Learning," which lists the following food-related inventors and inventions (alphabetically):

- Bar Code, also called Universal Product Code or UPC, which was first developed and patented in the 1950s by inventors N. Joe Woodland, Bernard Silver, and David J. Collins.
- Bread Slicer by Otto F. Rohwedder and Gustav Papendick in 1928.
- Burbank potato (or Russet Burbank), bred by Luther Burbank in 1871.
- Can and can opener invented in 1810 by Peter Durand.
- Carbonated water first produced by Joseph Priestly in 1767; first carbonation device invented by Torbern Bergman in 1770.
- Chocolate chips invented by Ruth Wakefield in 1930.
- Coca Cola, first invented by John S. Pemberton in 1886.
- Cotton candy soft confection invented by William Morrison and John C. Wharton in 1897.
- Dutch chocolate (rather "dutched" or alkalized chocolate) was invented by Coenraad J. Van Houten in 1828.
- Hot Dogs first sold by German immigrants in New York City in the 1860s.
- Kool Aid (powdered drink) invented by Edwin Perkins in 1927.
- Life Savers peppermint candies invented by Clarence Crane in 1912.
- Marshmallow was already being made by ancient Egyptians over 3000 years ago.
- Mayonnaise probably invented in 1756 by a French chef working for the Duke de Richelieu.
- Mechanical reaper for improved harvesting invented by Cyrus Hall McCormick in the mid 1800s.
- Microwave oven invented by Percy LeBaron Spencer, first commercialized in 1954.

- Pasteurization invented by Louis Pasteur in the mid 1800s.
- Popsicle invented by Frank Epperson (at the age of 11!) in 1905.
- Peanut-based products first developed by George W. Carver around the turn of the nineteenth to the twentieth centuries.
- Potato chips invented by George Crum in 1853.
- Refrigerator had several periods of invention; first in 1748 by William Cullen, followed by Michael Faraday in the 1800s, Oliver Evans in 1805, John Gorrie in 1844. First electrical refrigerator invented by Thomas Moore in 1803; first commercial refrigerator sold by the General Electric Company in 1911.
- Sandwich invented by John Montagu, the 4th Earl of Sandwich around 1762.
- Sugar processing patented in 1864.
- Teabag invented by Thomas Sullivan around 1908.

(http://www.enchantedlearning.com/inventors/food.shtml)

Figure 8.2 shows a drawing of the principal cooling vessel from Thomas Moore's "Refrigerator" patent of 1803.

Many devices were invented in the area of fermentation and finishing off fermented products, for instance by distillation. Figure 8.3 shows a drawing of a cider distillation device from 1864.

This list is of course incomplete and does not show some of the more recent food and beverage innovations that I have briefly mentioned earlier on and in previous chapters (hypoallergenic baby formula, instant coffee with all the difficulties of balancing extraction yields with aroma and taste, coffee in capsules, purified and re-mineralized drinking water) and others, such as the introduction of pro- and pre-biotics into appropriate food carriers such as yogurts and other dairy products, as well as the entire area of biotechnology and biotransformation, be it in the beer or wine industry or "green chemistry" applications in foods in general, more specifically in the areas of nutritional, textural, or flavoring ingredients.

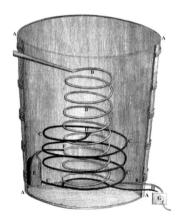

Figure 8.2 The cooling vessel depicted in Thomas Moore's patent of 1803. (Reproduced with permission of Google.)

Figure 8.3 Cider disitillation device. (From Jakob Schnyder, Uttewil, Switzerland, 1864).

WHERE AND WHEN DOES INNOVATION BEGIN?

Earlier on, I stated that innovation cannot really be taught and that it is almost a character trait that we are born with. Said in this way, it is most likely a false statement, in as much as one should say that at a certain age, there is a certain degree of maturity and past experience, and innovativeness has either become an important character trait of a person or not; in the latter case, and that is what the statement really means, innovation and innovativeness cannot really be taught anymore. It could simply be too late, and yet, one should exclude nothing, everything might be possible, it's just getting more and more difficult and increasingly unlikely that the capacity of being innovative can successfully be "grafted" onto such a person.

So, when does this all start? When is the right moment to learn creativity and innovativeness? I would suggest that this begins at a very early age, in our families and the ways that little children are being challenged when it comes to letting them find "their way around" or, alternatively, how parents show the way. This is not a book about education and formal learning, but I dare make the statement that the more still very young children are allowed to discover their own ways to the solution of problems and issues—all this in a protected family or peer environment—the more such kids will become creative problem-solvers and will more likely end up in the group of creative and innovative people that we have described throughout this chapter.

The short answer to the question is therefore: innovation has to start very early in life, optimally even before Kindergarten, within the family. From experience and observation it should also be said that creative parents are not necessarily always an asset, as they might be so creative and full of the desire to solve problems themselves that they do not give their children enough freedom to find answers and solutions of their own. Creative and innovative parents should therefore be rather hands-off and let their kids do the discovery, and not so creative parents should probably just be patient and be in awe of the creative minds that their kids develop all along. Well, that's enough on education and children's behavior, but I feel that is an important, in the eyes of some, maybe even trivial, observation. The good thing is, because it is trivial it should not be too difficult to let it happen. Once young and not so young kids are on their way to creativity and innovation, and without any bad interference from too formal an education, it is very likely that such kids will always be creative and innovative.

The late Nicolas Hayek, one of the co-inventors of the Swatch® watches and co-founder of the Swatch Group, was interviewed at the occasion of his 80th birthday in 2008, two years before he passed away, and was asked, not surprisingly, what was the secret of his success? Mr. Hayek said the following (and these are the words as I have them memorized and not an exact transcript):

> *When I was six years old, I was so curious and at the same time naïve, as a six-year-old would be; I wanted to know everything and asked all the naïve questions that a six-year-old would ask. When I was 16, I had added 10 years of experience to the six-year-old inside me, which still wanted to know everything and asked the naïve questions of the six-year-old, but wouldn't take the same answers anymore. And so, throughout my entire life, I have always kept the naïve and curious six-year-old inside me and added the years of experience to it.*

Mr. Hayek was most certainly an extremely successful innovator, creator and, executor, who, together with his colleagues and the people of the Swatch company not only turned around a dying industry in the early 1980s, but was at the helm of the biggest watch company in the world when he passed away in 2010.

The real question is: did you keep your six-year-old inside you? Do you actually dare to admit that you still have him or her inside you and that he or she (or let me say it, it is still rather young …) asks all these questions and wants to know everything about everything that you feel is of importance to it/you? It is well understood that you do not want to appear naïve in front of your peers and bosses, but I strongly recommend that you nurture and pamper your very own six-year-old, deep inside yourself!

The real answer, or rather recommendation, is finding the right balance between the apparent naïveté of your inner six-year-old and the street-smart and efficient person that you want to appear and be in your work environment. This is certainly not trivial and easy to achieve, but, from experience, I can confirm that it actually works pretty well and is easier done than thought. You may try it out with your family and friends first and then, why not?, you may dare to apply it in the workplace. Most certainly you will see successes with such an approach and will learn to become very good at it. Go for it!

THE PEOPLE IN THE FOOD INDUSTRY

Quite a few years ago I had the opportunity to listen to a speech by the CEO of a large food company about the kind of people that the food industry, his company more precisely, would typically hire. What he said was rather surprising: "Our company does not want to hire the best, but the most reliable; they often might only be the second best." Wow, that was an eye opener and I had to digest it first. Why would he say that? I had thought that it is the clear intention of every company to get the best people working for them, is it not? Sounded like it isn't. The most reliable would mean, in my eyes, the most grounded ones, the most solid ones, the ones that do their job, probably above average, but not necessarily brilliantly. This statement still echoes in my head and I am still wondering about its value and usefulness for the industry. When looking at other industries and from experience based on many interactions with colleagues in other industries, especially, for instance, in Silicon Valley, and even more specifically within design firms, their very outspoken goal is to hire the best: apparently there are enough of those best, so the company can choose and select the one who is "the very best." So, how does this fit together, hiring "the best" in one industry and hiring the "most reliable" people in another? At first sight, it does not fit and it almost might appear as if the food industry puts itself at a disadvantage to other industries in the area of consumer goods or fast-moving consumer goods.

Let us look in a bit more detail at this apparent contradiction, or at least strange phenomenon. The first questions we have to ask ourselves are: what's the definition of the most reliable ones and what is meant by the best? These are not trivial questions and, as one would expect, there are many different answers, some of which may even be surprising. In my opinion and also from many years of observation and having worked alongside hundreds of people, what really has to be high on the wish list of every company is neither "the best" nor the "most reliable," but the "most efficient" and fitting. This is a difficult task and in recent years many new evaluation, selection, and hiring procedures have been introduced in many companies. We went full swing from the lone hiring decision of a manager to a more inclusive evaluation process, where future potential colleagues of the candidate were involved to some degree in what I would call "hiring by committee." In the latter process, very formalized procedures are followed, many people are involved, sheets of paper, or IT-based control sheets are filled in, and it all looks very fail safe.

From experience I can say that at the very end it is still the highest-ranking person in the process who ultimately decides whether and who will be hired. Basically there is nothing wrong with such procedures, however they are rather lengthy and very impersonal. The latest fashion in the human resources department of many companies that I had the opportunity to work with in one way or another, is to not send any answers to applications anymore and people who have gone through the companies' websites, have found something they believed would match their profile, have duly filled in papers, have written clever motivational letters, and have their hopes high of finding a great job, most of the time will not even receive an acknowledgement letter. This is an unworthy fashion and can easily be corrected; that's what human resources are after all there for, at least one would think so. HR should help management in finding the most appropriate candidate, what I have called the "most efficient" one in an earlier paragraph. I have attempted to find convergence between the character traits of the best and the most reliable people and Figure 8.4 depicts this in a graphic way. The convergence goes to the top of the figure and along the lines I have listed the traits of the best (left half of the bottom line and left line going to the top) as well as the most reliable (right half of the bottom line and right line going to the top). The idea is to show that the higher the traits are towards the top, the more important they become, from both sides, leading to a mixture of required traits that would characterize the ideal candidate as the "most efficient."

It can be seen from this figure that I really put networking capabilities at the very top of this triangle, which I believe very firmly is probably the most important of all the necessary characteristics of a person working in the food industry today. In today's corporate

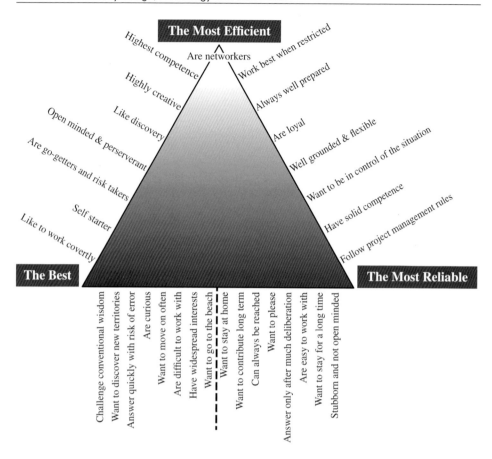

Figure 8.4   Convergence of character traits.

environment, networking has really become an absolutely paramount quality for any employee, inside as well as outside the company. I can truthfully say that the slogan "Network or Perish" is of ultimate importance. In sync with this emphasis on networking, Chapter 12 deals exclusively with this topic.

Many of the other traits are pretty obvious and I have listed them based on observation and personal experience. Let me add one additional thought: it could be that the above mix between the best and the most reliable can also be achieved by hiring two different types of candidate, such as the very fresh and rather inexperienced college or apprenticeship graduates, as well as more mature and more experienced and seasoned employees. I am totally aware that in many of the readers' work environments you may find variations on this theme, as I am also aware that one size, one way, does not fit all needs. I encourage the reader to perform a similar exercise and find out what appears to be different in their industry and which model, if any would be more efficient and work better. I would especially respectfully suggest that representatives of HR departments give it a critical look and, if appropriate, adapt it to the specific needs of their company. I can say with some conviction that today's hiring procedures in large corporations such as food companies are sub-optimal and two critical elements appear to be missing: respect for the candidate and the guts to take decisions without asking an entire "committee" and requiring a lot of IT and other tools.

COMMITMENT TO INNOVATION

There is one more aspect that we need to briefly look into in this section, namely real commitment to innovation. Often, employees deplore their environment and their boss, who does not let them be as creative and innovative as they could possibly be. I can rather directly say that most often this is simply not true, without wanting to use a stronger, more descriptive word for this situation. Chapter 9 will discuss the theme of "Nurturing the Innovators" in more detail. However, in this section, I want to express a few thoughts regarding this seemingly difficult situation for the individual, who has all the drive for creativity and innovation and is simply frustrated because of this feeling of helplessness or incapability. This feeling is real, at least it is perceived as real. And it becomes stronger and even more real, the further one is away in the value chain (or rather the "value maze") from the end point, namely the consumer. R&D representatives are probably the most frustrated in the entire chain, and often disappointed that their ideas, their initiatives, and their concepts did not make it to the very end. It's a normal feeling, because there are still so many imponderables in the chain of events that lie ahead and so many things can go wrong. I would say, again based on personal experience, that those who work in the R&D department of a company are probably the ones who are exposed most to rejection and subsequent frustration and, over time, get used to it. This getting "used to it" is, however, very dangerous and counter-productive, and is a serious value destroyer. Instead of "climbing the barricades," people in large organizations have a tendency to blend in and do not want to appear too much of a person that stands up for what he or she believes, but, over time, with experience and frequent exposure to frustrations will just go with the flow. And, admittedly, it is not easy to break out. This is the moment when some of the traits described above in Figure 8.4 come into play.

Take, for instance, the traits such as "go-getter," or "like to work covertly," "are always well prepared," are "well grounded" or "challenge conventional wisdom." The fear that most would have is the fear of being punished by the hierarchy, often not even immediately, but slowly and steadily, by just not getting this promotion or being taken off a project or sent off. These are real dangers and must not be neglected. However, creativity and innovation again can play a major role here, together with one important feature that can definitely be learned and which, quite purposefully, I have not listed amongst the traits. It's "street smartness," the kind of smartness that is a mix of intelligence and survival skills in all their facets: knowing how to play the system, always being respectful yet determined, anticipating what might come next and being a step ahead, ultimately becoming strong enough not only to digest frequent frustrations and disappointments, but actually turning them into something constructive, effective, and successful.

I have put much emphasis on people from the R&D levels, however, similar situations of frustration can be experienced by anyone in the entire chain of events, to a greater or lesser degree. The main point is just to say that the further away you are from the end point, the consumer, the more likely it is that you will experience such situations of being rejected and becoming disappointed.

There is a famous and often quoted line from a song composed by Jerome Kerns and Dorothy Fields in 1936: "I pick myself up, dust myself off, start all over again!" This should really become your Leitmotiv, which can show you the way out of difficult situations. I dare say that creative and innovative people might have an advantage when it comes to "picking themselves up ... " as they might more easily be able to find creative solutions than less creative and innovative people. So, "picking yourself up, ... , starting all over again," is a really strong show of commitment to creativity and innovation, invention, and, ultimately, re-invention of oneself, over and over again.

CONCLUSIONS

In Chapter 8 we discussed and analyzed topics of creativity and innovation and how this is specifically linked to value creation. People and their attitudes and how these are linked to creativity and innovation were discussed and the topic of being an innovator in the food industry was discussed and analyzed in some detail. This was followed by a very short historic overview of inventions in food and beverages and the question of when in a person's life creation and the desire to be innovative begins was discussed.

We looked at the topic of people and their character traits and strong personal points in the food industry and terminated this chapter with a short discourse on people's personal commitment to innovation:

- We introduced the "four stages of value creation," in which creativity and innovation can be seen as the starting points, followed by the equally important stages of creation, i.e. the translation into concepts and concrete examples, and finally execution or practical realization in the form of products and processes.
- The next important topic we discussed was about people and their attitudes towards the four stages of value creation, especially towards creativity and innovation. We discussed the question whether anyone can be creative and become an innovator, and whether creativity is a character trait that we are born with or whether this can be acquired. We made a short excursion to some basic neurological specialties and how, in the view of some experts, the brain appears to function, and introduced the idea of the strong interaction and balance between emotion and reason.

    We asked the question of whether the capacity to be an innovator can be learned or taught, and elaborated on this in a later section.
- We discussed the question of how to be an innovator in the food industry and found that in one important ranking with the title "Innovation Leadership Main Factors Behind Top Companies' Success", six out of the top 20 companies listed were purely food- or beverage-related enterprises.
- In order to put "some meat to the bone" we reported on two studies that dealt with inventions and innovation in the food and beverage industry. The interesting part that "refrigeration" and the refrigerator was listed as the No. 1 invention in one study and was also mentioned as a key invention in another. Interestingly enough, the first patent describing a "refrigerator" dated back to 1803, so more than 200 years back.
- We briefly touched upon the question of when innovation begins in a person's life and whether the capacity to innovate can be taught and be acquired by the individual. We introduced the example of Mr. Hayek's six-year-old, the six-year-old in every one of

us, who is curious and naïve and wants to know everything. We suggested that if one starts early, one can learn to become an innovator.

- We then discussed the topic of people in the food industry and what types of personalities are preferably hired or are preferably interested in working in the food industry. We drew the "triangle of character traits" with "the best", "the most reliable," and ultimately "the most efficient," the most suitable for the industry. The food industry does not simply hire "the best," but a suitable mix between the best and the most reliable, namely "the most efficient" for the industry.
- We finally concluded with a short topic on commitment to innovation and that very often people's disappointments and frustrations are not necessarily caused by the hampering influence of the hierarchy ("My boss did not let me"), but are more often than not self-inflicted. "I pick myself up, dust myself off, start all over again!" is the new Leitmotiv for disappointed innovators.

TOPICS FOR FURTHER DISCUSSION

- Discuss and critically analyze value creation in your company, and the influence of creativity and innovation. Are important stages missing, and if yes, which ones?
- What is your personal experience and history of creativity and innovation? Do you consider yourself to be innovative? At what level on a scale from 1 to 10?
- Discuss and analyze examples of "late innovation bloomers" and analyze the pathway of how they got there, assuming that you can find such examples in your immediate or more distant work environment.
- Discuss and analyze your company's overall innovation capacity, as well as desire, for innovation (not on paper, but in lived reality and examples). Are you happy with the situation or could it be improved? Does the company even do too much and cannot keep up? How could you personally contribute?
- Discover and list even more food- and beverage-related inventions that are important in your perception and which were not listed in this chapter.
- Discuss the question of whether innovation can be taught (or acquired through learning for that matter). List and discuss arguments for and against it. What is your personal experience?
- Do you have your own six-year-old inside yourself? If yes, how does it express itself? Describe your experience.
- Critically discuss and analyze the "triangle of convergence of character traits" of people in the food industry. Are you in agreement, based on your experience and given the environment you work in? What, in your eyes, is missing? What is obsolete?
- Discuss and analyze your own, personal commitment to innovation. Do you see more hurdles or more opportunities? When was the last time you "picked yourself up?"

REFERENCES

BBC (2007). Top Gear, *Series* 10.
R.E. Cytowic (1993). *The Man Who Tasted Shapes*, G.P. Putnam & Sons, New York.
The Nation Multimedia (2012). http://www.nationmultimedia.com/business/Innovation-leadership-main-factors-behind-top-comp-30183428.html

9 Nurturing the innovators

You can dream, create, design and build the most wonderful place in the world … but it requires people to make the dream a reality.

Walt. E. Disney

"PEOPLE ARE OUR MOST IMPORTANT ASSETS"

The concept of "the right people" has already been very briefly introduced in Chapter 5. You must have heard "people are our most important assets" many times in motivational talks given by your management or representatives of your human resources (HR) departments, and every time you may have answered to yourself: "Yeah, exactly, but I can't see it." You may have had doubts that it was really meant the way it was said, and your suspicion is that it's rather a lip service statement. On the other hand, nothing works without people, after all. So, even if the statement sounds a bit motivationally tainted, it has a strong true meaning. However, I dare say, it's only half of the truth, as the statement as such is not complete.

We have discussed in great detail the most desirable character traits of people that work in the food industry and have suggested that a convergence of "the best" and "the most reliable" to give "the most efficient" is the preferred personality to hire into the food industry. In other words, these most efficient people are the right people, the ones that will move the company forward and make it even more successful than before. The above statement should therefore really be as follows: "The right people are our most important assets." That sounds trivial and makes a lot of sense, but gets often forgotten in HR good-speak. You really should make this your mantra, namely that you are one of the right people and you, as the right one for the job, are the most efficient and therefore the most important asset.

Management in every industry, especially in the food and beverage industry, must realize that it's this large group of "right people" that makes or breaks the successes of any corporation and that these right people have to be well taken care of, and should be provided with the best possible conditions in their work environment, and should also be appropriately recognized, in whatever form, both individually and as a group, for all the successes they have and will bring to the company. This is probably the most important task of any management: provide the right environment, and recognize and praise. I know this sounds simple,

Food Industry Design, Technology and Innovation, First Edition.
Helmut Traitler, Birgit Coleman and Karen Hofmann.
© 2015 John Wiley & Sons, Inc. Published 2015 by John Wiley & Sons, Inc.

and actually is simple. From personal experience and looking into my own work situation over many years, I can say that I went through many phases of both, first the evolution and development of my own personality within a large group of colleagues and co-workers, and secondly the various phases of having been provided, or alternatively not been provided, with the best possible work environment and having had recognition (or not) and praise (or not). I must also say that there is a lot of subjective perception in all this, and I was often astonished to realize that what I appreciated as great praise by my boss was perceived very differently by my colleague who had worked together with me on the same project. This different perception has a lot to do with different degrees of expectations and probably also with the "years of service" in the company and therefore having been conditioned through work experience in different ways. I discuss this in some detail, because I want to demonstrate how difficult it can be to provide to and nurture a group of employees who collaborate on an important project, because each member of the project team may have a completely different level of expectation. This makes managing such groups rather tricky and totally non-trivial: it will have to become an exercise in recognizing the individual personalities in each group, finding commonalities and applying one's best management skills.

You probably have already harmlessly joked about the various management styles in your work environment and we can list a few of these here: management by delegation, management by committee (the boss always involves a number of people to take decisions and deal with situations), management by fear and intimidation (very popular these days in times of relatively high unemployment rates and difficulties in finding jobs easily), management by default (whatever happens is right), management by hierarchy (not the same as "by fear," "I am the boss"), management by "divide and rule," management by anticipation (more specifically "anticipatory obedience" to the manager's boss: "I am an even better soldier"), management by hope ("I hope they know what they do, so I let them do"), management by absence (similar to the hope version), management by following all the rules, management by inventing more rules, management by e-mail, management by post-it-notes, management by meeting (it is believed that work is only accomplished in meetings), management by power point ("showing even more presentations will explain what I expect from my people"), management by proxy (I ask HR to solve my management challenges), management by buddies (I have a preferred group of people through which I communicate to the other members of the group), and maybe a few more that I have forgotten here. I invite you to come up with your own list. It is important to recognize what is going on in your work environment and how you may best react to it. But, if all these management variants and styles are not the right ones, what is implied in all this; what is the right management style? In which environment would you feel best nurtured and looked after, guided, and the best brought out in you? Strangely enough, I strongly believe that there are many features in the list above that can be thoroughly helpful in crystallizing this perfect management style, the one that does it all. I suggest that it does not exist, and that's the whole difficulty. Every manager has to find his or her style, best adapted to the group that is managed, to the situation, the status of the project, the individuals in the group, even seasonalities (for global companies organizing a meeting in Singapore on, for instance, July 4th and inviting US participants is rather a no go, yet often forgotten).

From experience, observation and having lived so many different situations, I would propose that there is one management style that over-rides all the others, and, in all situations is always the right one: management by respect! Try it out, it will always work. If you are on the "receiving end," ask for it, request that your manager manages by respect, if your manager does not do it already. It may not be easy to ask, and may sometimes even be as painful as a root canal, but, very much like the root canal, you will feel much better afterwards. Guaranteed!

HOW THE RIGHT PEOPLE ARE BEST SUPPORTED: DEFINE VALUES

If you don't water a plant, it will simply stop growing and eventually die. It may push a few blossoms before it does, but over time it will still perish. It's pretty much the same with people; if you do not nurture people, meaning that you give them challenging opportunities, tell them what is expected, and then follow them from a slight distance, they will eventually lose interest and you will have lost them. Winning them back is more difficult and costlier than treating them right from the start. One of my former colleagues gave me the following slogan, which he had "inherited" from his father under the title "From the Memories of Harold K. Roth—Hal's Helpful Hints for a Happy Workplace":

- Hire people whose personalities are compatible with your own
- Pay them fairly
- Train them thoroughly
- Get out of their way and let them do the job for which they were hired.

(T. Roth, Personal Communication, 1996)

It's a very simple wisdom and so often seems to be so difficult to apply. If I would add anything to those four lines it would be just this one: from time to time check results and give constructive feedback.

If my co-author Birgit would add anything it would be the following:

- Make sure resources are available to give him/her time and ability to pursue a "to-explore list" on top of his/her "to-do list," meaning a "to-do list" should never take over 100% of a day, instead there should be time (up to 20%, see the Google 80/20 Innovation Rule in Chapter 8) for the "to-explore list," such as attending or speaking at conferences and (networking) events, pursuing side projects, validating ideas through various meetings, and with a budget for market studies or prototyping, etc.
- Now, we need to take this conversation to the next level and include Gen X and Gen Y. Generation X, commonly abbreviated to Gen X, is the generation born after the Western post-World War II baby boom describing a generational change from the Baby Boomers. Demographers, historians and commentators use birth dates from the early 1960s to the early 1980s. Generation Y, also known as the Millennial Generation, is the demographic cohort following Gen X. There are no precise dates for when Gen Y starts and ends. Commentators use birth dates from the early 1980s to the early 2000s. Both generations will have different work and learning/training conditions and requirements from the Baby Boomers (source: Wikipedia).
- There are some tech companies in the San Francisco Bay Area from which we can take away some innovative approaches in terms of working conditions catering for these new generations. Let's explore Netflix's unlimited employee vacation policy and

why it works. Netflix is the world's leading internet television network and it has one less thing to keep track of: the amount of vacation days their employees take. Netflix has a "freedom and responsibility culture," followed by IBM with their famously flexible time-off policy—letting employees leave early and take a day off on short notice, and also various start-ups and innovative companies working in the tech space are beginning to see the benefits of replacing vacation limits by promoting personal responsibility. Now, we have to ask ourselves the question, how we can translate and "copy/paste" these tech examples into the food industry?

- Gen X and Y learn differently, use different tools and also communicate differently. Salman Khan actually started to breathe some design thinking into the traditional education and training field by "reversing the classroom," also called the "flip teaching" approach. Flip teaching is a form of blended learning in which students learn new content online by watching video lectures, usually at home, and what used to be homework (assigned problems) is now done in class with the teacher offering more personalized guidance and interaction with students, instead of lecturing (source Wikipedia).
- If you want to reach Gen X and Y, you had better know "Snapchat" or "Instagram" and find a middle way to your modus operandi called email. If you want to retain your best employees, you also have to understand how they "tick" and what matters to them. Gen X and Y want to be part of a cool, fun, and engaged community. It is common for tech companies in the San Francisco Bay Area to host so-called "town-hall meetings" every Friday. There, the CEO of a company, be it Mark Zuckerberg of Facebook or Jack Dorsey of Square, etc. welcomes new employees, shares insights and challenges about the business, invites guest speakers (for example, Facebook had President Barack Obama at their town hall meeting in April 2011), followed by a Q&A (question and answer) session. After Q&A people mingle and network over a glass of wine or beer and snacks, while a DJ or band plays, which makes "networking" within the company fun and makes employees feel part of it, but it is also less hierarchical—like being able to see and interact with your CEO on a regular basis.

Treating people right also means that, if necessary, they have to be criticized and corrected, and they have to learn to deal with it. The worst situation for everyone, employees and bosses alike, is indifference. This has to be avoided at all costs and it is important to realize that both parties, the "right people" and management, have a role to play in this. Let me develop a few suggestions of best practices when it comes to making and maintaining the work environment as exciting, attractive, rewarding, efficient, sustainable, filled with riddles to solve, and opportunities to deliver great results over and over again. Everyone should have this feeling, once so well expressed by Chinese philosopher Confucius: "Choose a job you love, and you will never have to work a day in your life." The love for the job does not come on its own or is automatically based on one's own intentions and expectations, but has to be reflected and expressed through the work environment, through colleagues, peers, hierarchy, and ultimately by the possibility of everyone in the organization to be as close as possible to the consumers as the ultimate target for all the effort put into successful projects and processes. Everyone has to work together towards the same goal, and that is probably a rather difficult undertaking. Because people are involved, even if we assume that they are all "the right people," egos are equally involved and, despite the best efforts of many, individual and personal targets are not necessarily in sync with each other.

One important first step to harmonize different individual interests is to develop a list of common values that are not only quite obviously in line with the company's overall values and ethical, management, and leadership principles, but are also jointly discovered, discussed, adopted, and lived on a daily basis.

It is important to emphasize at this point that the values I am discussing here are not of the same nature as the "value" in value chain or value maze, as discussed in Chapter 6. Whereas value in Chapter 6 refers to financial value, value in the following discussion means beliefs, goals to achieve, or simply character traits.

Without such agreed upon values no one group can successfully operate towards a common goal and perform in a sustainable and successful fashion. This might sound trivial, but from many years of personal experience I have to emphasize the importance of this act: despite the belief of the group that they know the company values and act accordingly, it is paramount to sit down with the group to discuss and re-discuss such values and, most often, add additional ones that are discovered to be of importance to the wellbeing and success of the group.

Figure 9.1 describes a set of values, structured into tangible versus intangible values, and expected versus unexpected. Tangible means that adopting and pursuing such values is of direct, concrete value to the company, whereas intangible means there is indirect, not immediate, value for the company. Expected means what it says, namely unsurprising, whereas unexpected values are unforeseen, may be harder to understand, but are of great

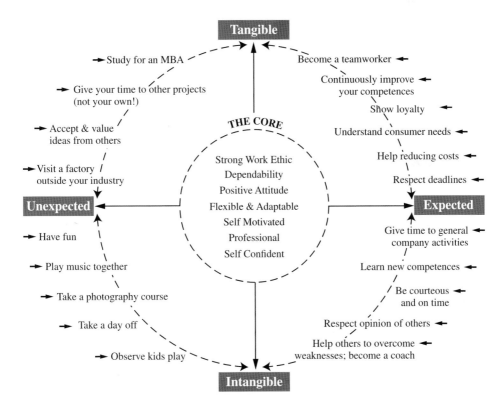

Figure 9.1 The "Wheel of Values."

potential to the company's wellbeing. In the center of all such values we find a set of core values that may or may not be complete in your eyes, and I would strongly urge you to discuss both, the core as well as all the others, with your colleagues and peers in detail. You might be surprised how little agreement you find in such discussions.

I have chosen these values based on own experience, but also on what typically can be expected when defining values in the workplace. Let me just look at a few of the values above, starting with the core values. I might even call them the "very much expected," values that sit at the core of every person and team that operates in a corporate environment. Some of these are of general importance and can equally be applied and lived in your family or amongst your friends. These are values such as: dependability, positive attitude, flexibility, or self-confidence. Even a value such as strong work ethic may have its place in a family environment, as it points to the fact that you are ready to put your "money where your mouth is" and do things for the greater benefit of your family. Probably of even greater interest may be the values that I have placed around the core, namely the tangible, intangible, expected, and unexpected sets of values.

I have also tried to give these values some specific weights depending on their individual locations in the four quadrants. A value that sits high up, e.g. NNE, such as "Become a teamworker" weighs heavily on the tangible scale and is something that can be expected from anyone in a large organization that is based on projects, their execution, and people working together. Another example of the "far East" on the quadrant would be "Give time to general company activities"; this is a value that is totally intangible, and which I would count as an expected one. An example of this could be spending some time in the company's internal fire brigade or being an organizer in the company's sports clubs. The value "Have fun" is both unexpected and intangible. There may only be a few people in your hierarchy who would accept "having fun" as an important value in the organization, or "playing music together" as anything of importance, hence they are intangible and would not necessarily be expected to show up on any list of values, and therefore very unexpected. And yet, and yet, from long years of personal experience and observation of how innovators work, individually or in teams, I can safely say that the southwest quadrant of this Wheel of Values is maybe of greater importance than one might think, maybe even of greater importance than some of the other values shown in Figure 9.1.

This is a very important realization, namely that seemingly unimportant or even unrelated things, especially values, may be of much greater impact on the functioning and wellbeing of individuals or teams, especially when it comes to innovators. I do strongly encourage the reader to take this Wheel of Values and use it as a starting point to establish, as a group, and together with your peers and bosses, if not already done, your company's, as well as your group's, your team's, set of values and use these as the absolute navigation system for how your organization intends to operate. Such defined values are after all the very basis for setting objectives and goals for the company, and, once agreed upon, also form the foundation of your company's successes.

Although I must assume that your company has done all or most of the above and has an agreed upon set of values, it is always very worthwhile to check this value system and critically question and discuss it every so often. The corporate environment changes more rapidly than ever before, and especially more rapidly than most of us, including your company's management, might want to admit. Or maybe they admit it, which makes it even more important to check and re-check your set of values, especially the ones that are not core, although even core values can and should be challenged from time to time. They might have become outdated, although it is not very likely.

Setting the right values in your organization and living by them is probably the single most important tool to best support the "right people" in your organization and help them to create a solid foundation for their creative and innovative work. Uncertainty in this important area is the biggest enemy of success.

CONTINUOUS LEARNING

Continuous improvement is an often heard and repeated slogan, and it has a very strong and critical importance in our work environment. Continuous improvement is the goal, continuous learning is the tool, and the desire to do so is the path towards the goal. Creative people and innovators have a natural tendency to learn more about what they do, many of them want to learn more about things that they do not do, at least at present, because they are typically self-motivated to enlarge the horizon of their knowledge base. What they have learned as students or apprentices is just a small part of what they want to learn and of what they are capable of still learning. This situation very much resembles the situation of top athletes: they want to become better and better every day, they have specific goals, personal or as a team, at given deadlines in the future (a new project for the one, a championship for the other) towards which they aspire; and, they ABSOLUTELY want to get there! Therefore, continuous improvement can be best supported and achieved by continuous learning.

There are three parts to learning: the offer, the opportunity, and the desire. It's as simple as this and it is as complicated as this. Your company should have, and in almost all cases does have, a fairly extensive offer of learning in the form of training modules in all areas: technical, business, finance etc. The offer is normally also very cleverly prepared and presented, and allows, in principle, everyone to participate without too many time and money constraints. And, based on personal observation, many employees go through such training modules and profit to the maximum of what they are able to achieve in one or two, or sometimes several weeks of training. Such training, in many large companies is carried out within the organization, through internal training courses, and is therefore full of typical company "good-speak." Critical minds may sometimes be turned off by too much of it, on the other hand, such training courses are an ideal vehicle to transmit company values and overall goals in a very condensed and efficient way. I would, however, suggest filling training courses additionally with some controversial, less unidirectional content, and by that ensuring that the critical minds, the more creative minds, the ones that want to "do it my way" are completely taken on board as well, find a forum for critical and constructive discussion, and through this maximize their learning. And it is not only the learning of the individual that would be maximized by this, but the critical individual's contributions to the company would become so much stronger and more sustainable because they are built on truthfulness and loyalty.

The major question that I would like to ask here is whether creative minds and innovators have specific learning needs that could be reflected in specific training tools. My quick and intuitive answer would be yes, they do need special training tools. It is like at school: special talents should be supported and encouraged in specific ways. I do not suggest that creative and innovative minds are better than anyone else, by no means, I just state the obvious: they are different and should therefore have different and well-adapted opportunities for training. Let me lean myself out a bit here by saying that the formal offerings in continuous learning that I have personally experienced and have been able to observe with many colleagues are, to say the least, often not really adapted, and sometimes even counter-productive. And yes, the more they cost, be it because of the time that was invested by the individual or because it was a specialized training course taught by an outside organization, the more they could be detracting from the main goal here: to give the creative and innovative minds the best tools so that they become perfect translators and executors of their creativity and innovativeness. Unfortunately, based on numerous observations, creative employees who are sent to management courses in the hopes of teaching them management, finances, or MBA-like attitudes, or individuals who want to get an MBA degree as a complementary competence, often come out totally frustrated for one reason only: nothing has changed, they still are in the same positions, they still have the same responsibilities, they just know a bit more about the business, without having the chance to apply it in a new function.

Yes, some are selected, but the level of frustration that is created in all the others is just too high, too costly, and too counter-productive for the individual, but most of all for the company. I believe a new way of teaching talents has to be developed and applied, based on individual needs, and which is also based on the promise of opportunity for each and every one who goes through such individualized, personalized training courses. We have to start clearly distinguishing between continuous learning for continuous improvement of competences and learning in order to take on new responsibilities and tasks. The latter, in many situations, is taken too lightly, people are sent there because it may be seen as a reward and not so much with the clear goal of: "Here is your new responsibility and you have achieved this through specialized learning." This would mean that fewer people would go through such, often very costly, training courses, and expectations can be met, frustrations will be minimized, and costs can rather impressively be reduced. This also implies the very important act of very serious pre-selection of those who should go to such specialized training courses, which is really the task of the management of the company. By sending fewer and better pre-selected employees to such courses, the money that will be saved can be spent by sending more creative talent to the "continuous competence improvement" training courses. I dare say that by doing so, companies improve the quality and outcome of ongoing training, reduce the level of frustration in many employees and, finally may reduce costs or are at worst cost neutral.

I do not want to dwell too much on the question of which training tools should be offered to the creative and innovative talents, but I want to suggest that much of this should actually come from them; they should be creative enough to develop their own, most appropriate training modules, more specifically the content, and then work with teaching professionals to create the best training tools. By this, creators and innovators have the chance to seriously contribute to their own fate, their own advancement, bringing their creativity to the fore and successfully driving it through the complex organization that typically surrounds them and to the very end, namely execution and to the consumer.

HOW CAN DESIGN CONTRIBUTE TO CONTINUOUS LEARNING?

I do strongly suggest that design in general and the various design disciplines can contribute in important ways to the continuous improvement and learning process in companies, especially in food companies.

By definition, training courses and teaching can and should be designed, by using several design disciplines such as media design, communication design, certainly graphic design, and, why not product design. I can say with almost certainty that training modules that are used in today's courses for employees, be they inside or outside the company, have never or only rarely been touched by any of these design disciplines. And yet, it would be so easy to make clever use of what is out there and not only adapt the content appropriately to the course participants but equally importantly, improve in really professional ways the way the content is put together, and communicated and taught. All this can be strongly supported by design.

Let me give a few, simple examples of what I mean. The most obvious could be the contribution from graphics design, but I would say that while the content is not optimally shaped and filled with meaningful information, it is no use embellishing it. However, product design combined with communication design could be the first steps. What is it that I want to talk about and teach to students in specific training courses? Can I conceptualize the theme, can I conceptualize the process or the product I want to speak about? For competence training, the contributions of media design and communications design could be crucial, as they might allow me to transform potentially boring training material into something more exciting, maybe even entertaining, captivating, and well formulated. The input from graphics design comes towards the end of this process; however, it might be important to involve the graphics experts early on, as a kind of guiding principle, and to give them a chance to accompany the process of creation of optimally designed teaching material from the very beginning.

Those of you that have already participated in one or more of your company's training courses, will have certainly seen that most of the course material is either printed or visual in the format of presentations such as power point. They all look very sleek and have most likely seen the hand of graphics designers or people who are good at that. I am, however, certain that behind this beautiful façade, no other design discipline had any input into any of the materials used in these courses. I suggest that you do a critical test of such material, as much as you can get your hands on, and check for yourselves if there is anything missing in terms of content, style, communication, and overall expected efficiency. I am not expecting you to be the training expert, after all, you are more likely than not at the receiving end; however, it would be totally cool to modify the material as you would see it fit, with a specific emphasis on design input. Ideally you would meet with design experts inside or outside the company and ask them for their advice. You will find such an exercise not only very refreshing, but extremely eye opening and will lead you to a whole new view when it comes to presentations, including your own, which, all of a sudden, almost become three-dimensional, not necessarily literally, but in the overall spirit of the material in question. As an additional exercise, I suggest you try and attempt a conceptualization in the form of a small model, prototype, or something that reflects the spirit of the presentation if at all possible. There may be situations where you will find this impossible, but it is safe to say that in many cases you should be able to do such conceptualization and prototyping of both the ideas embedded in a training course material or in your own presentations that you so hard try to render interesting to an audience. I will emphasize more on the role of design and designers in the ideation process, where not so much the "objective" part of design plays a role, but more the "process." This theme "from object to process" has already been discussed in detail in Chapter 2.

SUPPORTING INNOVATORS IN DRIVING THEIR INNOVATIONS THROUGH COMPLEX ORGANIZATIONS

Food companies are truly complex organizations. This sounds like the understatement of the year. In Chapter 6 I discussed the nature of the value chain and suggested that there is no such thing as a chain, but rather a maze, reflecting a very complex and intermingled connectivity between the various elements and financial values, as they would be traditionally described and looked at in the descriptor "value chain."

Figure 9.2 again uses the image of the "value maze" and overlays it with a possible pathway through a complex organization, a pathway that an innovator would have to find and follow in order to successfully drive his or her creative and innovative ideas, concepts, prototypes, and consumer offerings through a complex organization such as a food company.

I am totally aware that the pathway, depicted in Figure 9.2, which resembles more of a roller coaster than a straightforward march through the organization, could take many different connectors and sequences, and the line is to some degree randomly drawn; however, it starts at R&D and ends at the consumer, simply because, from many years of personal experience, more innovative ideas start in the R&D community of a company than anywhere else. It is also true, and was mentioned in earlier chapters, that more innovative ideas that originally started in R&D are buried than those that come from other parts of the company. It's a sad fact and has a lot to do with the complex and maze-like situation in large companies.

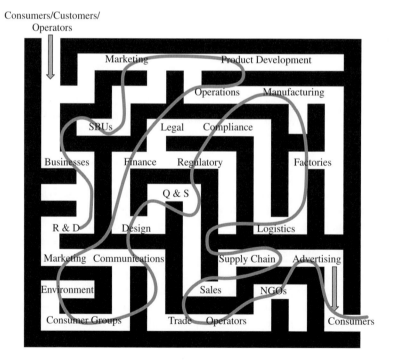

Figure 9.2 A roller coaster through the value maze.

What are the best ways forward to successful completion and launch of one's innovative ideas? Although there is no single and simple recipe, let me list a few straightforward and applicable suggestions. Let us begin with a series of questions that we need to ask in order to define where and how to situate any innovation idea and concept. The individual questions could be considered trivial; however, the whole list that follows is not. Most of the questions are fairly straightforward to answer and I suggest that you do some serious homework:

- Who are the internal business partners? How can you best involve them?
- Who are the main external partners for both development and launch of the innovation?
- Assuming that there is a stringent project management structure in place, is the project best embedded in the official structure or would you be better working around this ("skunk" work) and what is the risk in doing so?
- Who pays for the project? Normally this is less of a concern for people working in the R&D community; however, the question is an important one because the answer defines your supporter (mentor) or opponent.
- Whose support do you need to secure for your innovation and how can you make the supporters or the mentors tick?
- Duration, timing, and milestones are typical elements of project management; are they useful? How can you avoid distracting milestone and review meetings? Can the mentor help and show trust so that you can get away with going against the regular project management stream? Does it actually make sense?
- How much external resource, if any, do you need?
- Even if the innovation is still at a very conceptual stage, you should develop a basic business plan and think of how to sell the concept and, later on, the finished development. How do you do this and who should you tell the story to?
- Do you have experience in telling success stories? What does storytelling mean?
- How can you develop a captivating story board to tell your tale in the best flowing and convincing ways?
- Study and discuss key success factors of successful projects in your company. Do you know the history of such successes and what have you learnt from these, if at all?
- Do you have insight into potential stumbling blocks that will kill your innovation? Do you know the people who are potentially behind this? This is no call for conspiracy theories, but a simple fact: it is people who are behind failures; either oneself or someone working against you or the idea you represent.
- Maybe the most important question of all: how can you convince the properly identified decision-makers? This question implies that you know who these are and that you will eventually devise a strategy to not only get to them in person, but to have a chance to talk to them and convince them. The simple answer here is: dare! I know this sounds simple, and, from personal experience I can truthfully say that it is actually quite simple. The important part is not so much the daring part, but the knowing who to talk to!
- At the end of any innovative idea should be a successful execution and launch. How is success measured in your company? Do you contribute to the measuring process?
- How many success stories do you have to tell? Have you actually told them? Does your company provide a forum for telling such stories and, if not, how can you help create this?

- How much competence do you personally have in bridge-building, especially building bridges to important allies who can help you succeed in getting your innovation to a successful execution and launch?
- How experienced are you at balancing emotions versus facts in discussions? How do you typically reply to irrational behavior in discussions? As a hint, and based on many such irrational discussions, there is no use in replying with rationality; as difficult as it may seem, irrationality can only properly be responded by irrationality. However, hold your emotions in as much as you can.
- Finally, are you defensive in difficult discussions? Never (NEVER) be defensive, it is seen as a sign of weakness. Be determined and convincing instead. This will most likely lead you to the desired goal and to successfully driving your creative and innovative idea, concept, prototype, process through the complex organization of your company, and lets you prepare your next success story.

THE SECRET OF SHARING

Sharing, in large companies, is typically a counter-intuitive behavior. It's not what one typically does. Although we all know that, in life as well as at work, you will only receive when you give. So what does "giving" mean in a corporate environment? I have discussed some of this earlier in this chapter, when elaborating on the values, also depicted in Figure 9.1. Typical values that would support this giving attitude are "Give your time to other projects than your own," "Give time to general company activities," or "Help others to overcome weaknesses; become a coach. How can you share without losing out? First of all it means that you have to be self-confident enough (a core value in the "Wheel of Values") and possess enough street-smartness to properly judge what makes sense to share with others and what to potentially keep to yourself.

This sounds like "selective sharing" and in actual fact it really is. However, it takes nothing away from the concept of sharing and especially of the concept of "sharing is winning" (SiW). Let me introduce the basic rationale for SiW. We introduced SiW into Nestlé's activities around Open Innovation and the program of Innovation Partnerships. I will discuss this program as well as Open Innovation in the specific context of this book in detail in Chapter 11. However, SiW goes far beyond these topics of Open Innovation and Partnerships. With whom and what can you share and how much of it? These are not only questions of good behavior, and the respective answers are derived from the work environment itself. A first, very straightforward question that has to be answered is: who do you share your competence, your results, the information in general with? One would imagine the answer is: if it's inside the company it's with everyone, if it's outside the company, probably with no one. The real answer is more complex, however, and needs some more detailed discussion so that you get a better picture of who you may want to share with. The specific answer is always linked to the two other questions, namely what and how much should you share?

Let me give an example that may illustrate that in a simpler way. You and your team are working on a critical, "top priority" project and, despite your best efforts, you realize that you cannot meet the next milestone, in six weeks from now. You are stuck, you need help. It is quite natural that you go and see your boss or the person who is the receiving end in the business unit of your company and tell him or her all about your problem. That's the easy part, although it might be rather embarrassing and not so easy to admit. You have come to the conclusion that there is no way out and that you have to go and see the person and talk to him or her. What are some of the likely scenarios that you might be faced with? Traditionally and based on personal experience and obviously also very much dependant on the character of this reference person, you might be received very cooly, and not as understandingly as you may have hoped for. Do not be afraid, share the information that is necessary to make your point, clearly lay out the issue or issues at hand and ask for guidance, such as: would it be possible to extend the timeline, or could you obtain additional resources, or, and this may surprise you, is the project as important as it initially was and would a delay be at all critical? In other words, you do kind of a tit-for-tat: you give information that is crucial to understand so that the other person can offer appropriate, fitting, and agreeable ways out of the dilemma that you are facing. It's like with your report card in those school days: the only thing worse than showing the maybe not so good results, to your parents is not showing them. This is the same situation here: not talking to your peers about a possible issue with your top-priority project—or any other ongoing project for that matter—is the worst of all situations, always to be avoided. So, in true SiW fashion, this becomes the only viable way forward to resolve your issue.

There is probably no middle ground to be found here, it's either the whole truth or else. Up to this point, you may have followed my reasoning and are probably not only in sync with it, but have done nothing else in the past when you were in similar situations. The tricky part goes beyond the above two questions of what and how much, and is linked to the "how" can I best share what and how much? It's the most critical part of the entire equation: how do I go about this sharing? Do I do it right out, straightforwardly? Do I beat around the bush and approach it in incremental steps, thus creating suspense as well as impatience on the part of the person I am sharing with? Do I always have to do it in his or her office? Can I better succeed on my "home turf" or should I propose neutral ground such as the cafeteria, and share over lunch? The latter, however, is a bit tricky. One of the most over-used and under-delivered phrases in corporate speak is "Let's have lunch." It's the same as an Australian who bids farewell by saying "See you later," when she sees you off at Sydney airport on your flight to probably far away. This "see you later" is just a phrase and gives you the indication that it will maybe happen later, not specifying how much later. It's the same with "let's have lunch": it might happen one day or never, but it's so reassuring and comforting to express it and gives everyone a warm and hopeful feeling.

Coming back to the selection of the best location as well as occasion, from personal experience, the other person's office is both the most likely place, but also the worst place to have such a sharing moment, this "Come to Jesus" moment in the "lion's den." If at all possible, try to find neutral ground, don't plan it to formally, and come right out with what is the issue. Be respectful, but show that you are on top of the situation as it presents itself and, with all due respect, you ask for guidance and the best way out. Share as much as you can, but also only as much as is needed to resolve the issue. You will see, most of the time (not all of the time, it's virtually impossible!) you will win through such a sharing experience.

Let me discuss the other side of SiW, namely the sharing of whatever kind of information, data, know-how, or experience with someone from outside your own company. The first thing that is important to realize and to never forget is that information or knowledge is not only power, but also money, and therefore has a value that you have to know in order to be properly prepared for any sharing occasion. Like with the example above, when sharing you expect something in return; in the example you received (hopefully) understanding, guidance, and were shown ways to resolution. What is it that you expect in return when sharing anything with an outside party? I will discuss this question in more detail in Chapter 11 "From Open Innovation to Partnerships," as full understanding and appropriate application of this kind of bartering, including content and consequences, is the underlying foundation of the whole concept. At this point, I will just mention that in a SiW situation with an outside party you must never forget that you represent your company and you speak, think, negotiate, concede, and agree on your company's behalf, at least as much as your role in the company permits you to do so. Do not underestimate the power, but also the responsibility you have in such an exercise, so be very well prepared when it comes to knowing what, how much, to whom, and in which location. As far as the latter goes, the same counts as above: try to stay on your own turf or propose neutral ground; the other party will try to do so as well, so neutral ground is often the best. Such ground could be occasions such as conferences, trade shows, or specifically designed meetings at locations agreeable to both parties. None of this is really complex and almost goes without saying, yet I discuss it here, as we tend to forget these rules of the game, which are all based on common sense.

Let me add an innocent example to all this, which is far away from the food industry and, at first sight, has nothing to do with any of what I have discussed thus far on SiW. It's an example from the world of surfing. Some of you may have seen the *Endless Summer* movies (Figure 9.3). The first one is already 50 years old, the second one, with the rather unsurprising title of *Endless Summer 2*," was made 25 years after the first one. It's the story of two young surfers from Southern California, who travel around the world to find the best surf beaches with the perfect wave ("da wave") in places such as Hawaii, southern Africa, Australia and the Atlantic coast of Europe. *Endless Summer 2* expands to Indonesia and is an homage tracing the original trip. Gary Hoey's music is, to my ears, really great to listen to and extremely fitting for the movie. It must be said that the movie has no real story, but shows lifestyle and something much more important yet, namely "sharing is winning." How is that? Well, very simply, the two surfers do not stay alone in their surfing experiences, they meet many others, like-minded, and, especially in the second movie, very competitive surfers. They surf together, they train together, they show each other their material, they talk very openly about the latest improvements to their boards, and share the most advanced surfing techniques with each other, in every detail. And then, they go out and compete with each other and, not really surprisingly, there is always one who wins the competitions, and often it's the same one, the strongest, the best, maybe even the one who had shared most. There is probably no better and more fitting example of the SiW concept than in this world of surfing. I realize that there are not many sportspeople, if any, that show a similar camaraderie and sharing spirit than these surfers. This makes them so special and should be a good inspiration for applying SiW in your environment. I call it the "Endless Summer Syndrome," this desire to go out, meet other, like-minded people and share what you know, receive in return, put all this together and win, in your project, in your management ways, in your career, in your personal life.

PERSONAL NURTURING TOOLS

The following chapter "Useful Tools" will discuss the topic of the most popular and successful innovation tools in detail. This section, however, attempts to list a number of tools, helpers if you wish, that are important when nurturing the innovators around you or within you. There are probably as many opinions on the usefulness and efficiency of any of such

Figure 9.3 *Endless Summer*—movie poster. (With kind permission of Bruce Brown Films, Torrance, CA, USA. Copyright: Bruce Brown Films, 1966.)

tools as there are tools. Agreement or doubt and opposition to the tools is most often linked to whether you talk to management or to those who propose to use such innovation tools. Middle management typically is rather critical, although one has to look at this and discuss it in a more differentiated fashion. Middle management is torn between the desire to be in control, in charge of all that is happening in their sphere of influence, and on the other hand the need to deliver innovative solutions to their bosses and ultimately to the consumer. From experience, I always observed a kind of "go with the flow" mentality, or in other words: praise when the outcome is good and criticize when there are no immediate countable results. Basically, there is nothing wrong with this attitude and it is important to know for those who use innovation tools or are involved in any kind of innovation workshops that they should keep cool when they get pressure from their management. It's just part of the game and is totally ok.

In this section, I want to give a general introduction into a number of innovation tools, with real emphasis on innovation and creation, and shall not discuss tools for creativity. There is ample literature, which I suggest that the reader consults, if you want to dive deeper into this vast and complex topic. One of the better-known authors in this field is E. De Bono and he has greatly contributed to the field of creativity and unleashing one's creative potential. In the following link are listed a number of good and already standard reading materials and books on the topic of creativity (http://www.huffingtonpost.co.uk/2013/06/25/best-books-creativity-imagination_n_3495025.html).

Let me focus on listing and briefly discussing the most prominent and also well-proven innovation tools that are regularly used in the industry and have been used for quite some time already. I would group them in the following manner:

1. *Innovation tools or exercises based on joining together of "different and apparently irrelevant elements"* (e.g. Synectics; Gordon, 1961).
2. *Innovation tools or exercises based on situational stimulation of participants* (e.g. FastPack, IdeaStore; G. Lane, R. Lepior, M. Rodero, Nestlé, Personal Communication, 2000, Traitler and Saguy, 2009).
3. *Innovation tools or exercises based on consumer involvement* (e.g. IDEO, SRI International; Kelley and Littman, 2001, Carlson and Wilmot, 2006).

As listed above, there is ample literature, especially on groups 1 and 3, which I would suggest reading to gain further insight and be in a better position to judge the individual strengths and weaknesses of these innovation tools. One important consideration in all this is the ratio between costs and innovation output, which in some cases can become prohibitive. The other important consideration is to realize, in which status and environment such innovation tools are to be applied. And there is one more critical thought: you do not necessarily need consultants or external helpers to practice and apply any of the above innovation tools; with some practice and maybe just initial support, you should be able to run any of these creative problem-solving exercises that hide behind the three groups of innovation tools described above.

Moreover, it might be the case that you find more and different approaches to creative problem-solving and innovation exercises that do not fit into these three groups. I would be very keen to hear from you and learn about these. As mentioned earlier, the Chapter 10 will exclusively and extensively deal with the useful and undeniably successful innovation tools, largely focusing on the ones described in group 2 above, on which I can speak, describe, analyze, and discuss based on personal and frequent exposure to them.

Let me finally come back to the nurturing part of what I have briefly described and introduced. There are more reasons for people to participate in creative innovation exercises than to come up with doable and highly promising innovative concepts. There are at least two further important factors to consider: first, the feeling of the participant that he or she has been part of something important, something potentially game-changing and second, the very crucial learning to deliver under pressure and in a fast and expedient manner, both of which lead to increased self-confidence and the belief, no, the confidence that the individual has a strong role to play in the creative and innovative process. This is beyond "feel-good," this is of critical importance in any well-functioning company.

CONCLUSIONS

This chapter was mostly about people, values, continuous learning, and nurturing of the right people in a company. We also introduced the secret and special value of sharing and the important, yet still frightening step from being secretive about one's work to sharing more or less openly, yet smartly and by this I mean being able to create important networks, connections, and allies, all of which will ultimately help us to win in the struggle to become more innovative and successful.

- We discussed and analyzed the concept of "people" versus "the right people." We briefly discussed the role of management in looking after its people and the importance of "nurturing" the right people in an organization. The critical role of providing the right work environment for employees, and the contribution of constructive criticism and praise were discussed. We mentioned that this might all be trivial and yet, not always done in the right way and not always perceived in the same way by different persons on the receiving end. It may be trivial, but it's not easy.
- We introduced the idea, again rather obvious, that in order to nurture the right people and make them into winners, it is important that a set of jointly defined and agreed upon values has to form the basis of any successful and winning team. We introduced and discussed the "Wheel of Values," in which we grouped tangible, intangible, expected, and unexpected values around a rather basic set of general core values, the latter probably being very similar in every company.
- We discussed and analyzed the importance of continuous learning and continuous improvement, and we stated that there are basically three parts to learning: the offer, the opportunity and, last but not least, the desire by the individual to do so. We discussed the question of whether creative and innovative minds have specific learning needs and I suggested that yes, they might need specific, adapted tools. These should not necessarily be the popular and more glamorous training sessions such as MBA or management school, but really adapted to challenge creative and innovative minds. We also discussed the importance of follow-up to any training that the individual goes through and which is as much as possible in line with the expectations of the trainee.
- We asked the question of how design can contribute to continuous learning and especially training modules that are used in this. I suggested that design can play a much bigger role than it does today, namely helping make more or less fancy presentations

and that course organizers should really make use of the full offerings of today's design palette.

- We discussed and analyzed the very crucial aspect of driving innovation through complex organizations such as a food company and showed up the fuzzy pathways that innovators are confronted with, when they want to go all the way with their innovative concept and want to transform it into a winning product or service. I offered a series of succinct questions that the individual innovator or the team have to ask themselves and find satisfactory answers to, in order to become successful. The sheer number of questions shows not only the vast complexity, but more so the great difficulties in driving the innovation all the way through to the very end.
- We introduced, discussed, and analyzed the secret of sharing in large organizations, both inside and outside. This complex aspect will also be discussed in Chapter 11 in the context of open innovation and innovation partnerships and we suggested in this section that sharing has to be done in smart and controlled ways, tit-for-tat-like, yet daring enough that the returns are greater than what the individual gives away. We introduced the imagery of "The Endless Summer Syndrome," describing a maximum sharing attitude leading to maximum returns in winning.
- Finally we briefly introduced the concept of "personal nurturing tools," or in other words, useful and most of all successful innovation tools that are used to help creative and innovative minds to bring out their best and give them the feeling of being taken seriously, being nurtured, and being cared for. It is a typical, much-used "win—win" for the company as well as the individual innovator.

TOPICS FOR FURTHER DISCUSSION

- Discuss and critically evaluate the notion of "the right people." Do you have the feeling, or experience that the distinction between "people" and "the right people" is made in your company? If not, what can you contribute to ameliorate the situation?
- Have you personally ever participated in a discussion around values, be it your team's values or the company's values? Please take this point up within your team environment, and discuss and identify your team's values. This can either happen by listing what you already do and respect and by adding new, possibly surprising ones.
- Please discuss and define your own "helpful hints for a happy, efficient, respectful, fun, rewarding workplace."
- Critically look at the "Wheel of Values" and potentially change and adapt, based on your personal experience and work environment.
- Please list and discuss training courses that you have attended and analyze their post-event efficiency. Did you always go to the most logical or the most glorious ones? Did you ever dare to take totally unsuspected training courses (such as Steve Jobs' calligraphy course)?
- Please discuss how you would design the optimal training course in your immediate work environment. Try to incorporate design principles and find out the difference.
- Please discuss the questions in the section on supporting innovators in detail; bring the team together and take time to find the best and most promising answers.

- Ask yourself the questions: how much did I ever share? Was I rewarded for it or was I criticized for it? Did I ever experience the winning part? List the answers and discuss them with your immediate colleagues. Please share with your boss, as appropriate.
- Please list, discuss, and analyze the personal nurturing tools in your company, especially the ones that have to do with creative problem-solving and innovation.

REFERENCES

W.J.J. Gordon (1961). *Synectics: The Development of Creative Capacity,* Harper and Row, New York.

T. Kelley, J. Littman (2001). *The Art of Innovation,* Crown Business, Crown Publishing Group.

H. Traitler, S. Saguy (2009). Creating successful innovation partnerships, *Food Technology*, 03.09, www. ift.org.

C. R. Carlson, W. W, Wilmot (2006). *Innovation: The Five Disciplines for Creating What Customers Want*. Crown Business, Crown Publishing Group.

10 The innovation tools

The clearest way into the universe is through a forest wilderness.

John Muir

FROM RITUALS TO INNOVATION TOOLS

Why do creative and innovative minds need tools to bring the best out in them, especially winning concepts? There is no one simple answer to this question, but probably many hints that point in the right direction. I would argue that the basis of any innovation tool, or creativity and innovation support is rituals. Such rituals can be considered to be the precursors of the formal innovation tools. They are often seemingly stupid, and can come in many forms and shapes. I suggest three forms of such precursor rituals: spatial, irrational, and situational.

The first group of such rituals is linked to places and situations, such as taking a shower, walking in the forest, going to the beach, swimming, diving, driving in a car, playing ball, or performing sports and working out. These are *spatial rituals* and they are all associated with the simple, unifying fact of letting one's mind wonder in a free environment, not limited or narrow. Often the spatial dimension is linked to a dimension of movement, like when driving or performing sports. Some innovators have their most creative moments when they challenge themselves to the max, such as extreme workouts, running a marathon, or participating at an Ironman event. For most, however, simple situations such as the shower (almost a "womb moment") or meditating on the beach are sufficient to bring out the creative and innovative side of them, and can be considered as some of the best and most efficient rituals and innovation tool precursors.

The second group of such innovation tool precursors is the group of ticks or spleens, i.e. totally *irrational rituals*. Some innovators tell each other jokes to stimulate their creative juices, another has to wear a hat (always the same one!) when he or she sits down to perform a specific task, such as writing or composing or painting or creating new computer games or discussing an innovative marketing or sales plan. Others have to wear two different socks (red and green, for instance). They wear them purposefully, whilst their spouses find themselves at home with 19 odd socks … Yet others always have to wear their preferred sweater,

Food Industry Design, Technology and Innovation, First Edition.
Helmut Traitler, Birgit Coleman and Karen Hofmann.
© 2015 John Wiley & Sons, Inc. Published 2015 by John Wiley & Sons, Inc.

or T-shirt or slacks or sneakers, being certain that they make them better and more creative innovators. The strange thing, or maybe not so strange, is that all of this works; however, it is extremely personal and everyone has to find out for themselves what works best for them.

There is probably a third group of precursors, which I would call *situational rituals*. Innovators immerse themselves in specific situations, which have strong stimulating value for them. Well-known and proven situations are cooking a meal (for pleasure, not as a must), playing music (the guitar, ukulele, piano, etc.) or listening to music. All of these situations can lead to an almost meditational status and therefore are perfect innovation rituals. Some people become very efficient in this third type of ritual; they make repetitive and brief meditative situations a recurring theme during their work day and thereby can keep the creative energy always at a very elevated level.

Figure 10.1 shows in an overview the three different groups of rituals as innovation precursors. From the readers' personal experiences, there may be many more of such rituals that could fit into the chart, and I can only encourage you to find as many as possible of these, either within yourself or by observing and analyzing your environment, not necessarily only at work, but, often more interesting, with your family or friends. You will find out that there is a wealth of these rituals out there and they all work to varying degrees. Some are more efficient than others and, based on personal experience and observation, I would make the following top of my list in the three categories:

- Spatial: Take a shower
- Irrational: Wear a hat
- Situational: Cook a meal

I would not be astonished at all, if your personal hit list looked slightly different.

I have repeatedly stressed that these rituals are precursors for innovation tools and we will see the re-appearance of some of the above elements in the more formal innovation

Innovation Rituals as Precursors for Innovation Tools

Figure 10.1 Chart of precursor innovation rituals.

tools that I will present, discuss, and analyze later in this chapter. These rituals could be considered as a warm-up to the daily creativity, especially the situational rituals, which can be used anywhere and anytime, depending on their nature. You should define the ones that work best for you and apply them every day so that you become really good at them, basically in the same way as an athlete trains. Most of the ones described happen during our average day and can be used to become a wealth of inspirational moments and possibly innovative ideas. Even social networking, passively or actively, may be used for such moments, especially due to the fact that most of the time there are no really serious things going on, and reading or sending messages is not a highly intellectual activity, therefore leaving the mind enough space to wander and become inspired by the "nothingness" of the social network interactions. Certain network contents, reading or writing them, can be surprisingly inspiring, provided that you let your mind spin freely.

The rituals depicted in Figure 10.1, whether they are spatial, irrational, or situational, are all important ones, and the reader is strongly encouraged to use these occasions, almost free moments during a hectic day, to not just do nothing, but do nothing with meaning and purpose. This can become extremely effective with regards to training and improving your innovation skills, and subconsciously always being ready to accept creative and innovative ideas that come to you, without even knowing how and why they have arrived. There is, however, one more important additional element, and this is saving and documenting ideas that have welled up from your subconscious. There are probably as many means as there are people to document your ideas and ensure that they are not forgotten. The important thing is that you do it in such a way that you can easily retrieve them. One of the more popular and visual ways to do this is scrap-booking. This, however means that you have to have a scrap book always with you; in the shower this may be rather difficult. I am certain that you have your ways and I can only once more emphasize to be ruthless when it comes to documenting these ideas, even the apparently unimportant ones, as you never know whether in a given context they may not all of a sudden become very important.

There is a cute anecdote about the composer Franz Schubert. He made it a habit to use the cuffs of his shirt, which he always put on his nightstand, for his spontaneous ideas during the night when he woke up with yet another musical theme. He had ink and a quill pen always available and simply wrote on his cuffs. I do realize that we do not wear cuffs these days, we do not use ink and a quill pen either, but we still might wake up in the middle of the night with a great idea: be prepared!

THE INNOVATION ENVIRONMENT

A very important aspect of selecting the most appropriate innovation tools is the environment in which innovation is supposed to take place. What I mean by this is depicted in Figure 10.2. This figure nicely shows the different areas in which innovation is situated in the consumer goods industry, including the food industry.

The understanding of this chart is crucial to the success and the outcome of the application of any innovation tool. Simply put, it shows us the various possible environments in which any company can and should innovate, and how best to describe and define these environments. The two axes describe the situations of knowing the goal by knowing the consumers on the right side, versus the opposite, namely not knowing the goal on the left side. Towards the top we have the situation that we know the means, we possess technical

Mapping the Innovation Space

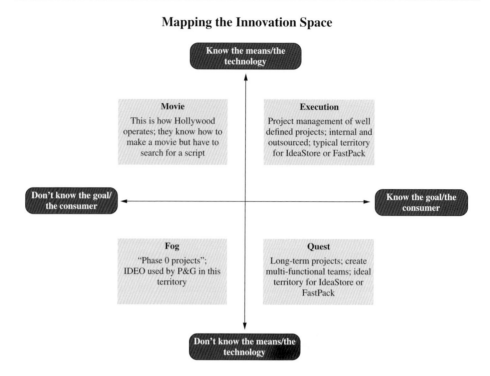

Figure 10.2 Mapping the innovation space (D. Sennheiser, formerly P&G, personal communication, 2004).

and other solutions, versus the opposite, we do not have any solutions yet, on the bottom of the chart. "Yet" is the important word here: we do hope to find answers through the creative and innovative process, helped by the most appropriate innovation tools. We thereby create the four quadrants, which I denote by their "geographical" orientations, i.e. north east (NE), south east (SE), north west (NW) and south west (SW).

Execution

Let me begin with the most obvious situation, namely NE, possibly the most likely to be found in your company environment, although you should not always trust all information that you get, either from the technical community: "We have the solution!" or the business community: "We know our target." However, should all this information and competence really be available, then one might think that we could be in "execution" mode in no time. There is, however, one important step to perform prior to execution, namely to find the right combinations, almost permutations of the possible goals, the most desirable goals, the most rewarding goals and the various technologies, procedures, ingredients, services, and general know-how at the disposal of an organization such as a food company, irrespective of whether it is small or large. This is not an easy task at all, it's actually extremely difficult and anyone who tells you to the contrary is either wrong or maybe has tons of experience in successful project execution with a proven track record. There are not too many of these around because they might just have been promoted too quickly and thereby have been removed from the talent pool of innovators.

In this NE sector, innovation tools like FastPack and the IdeaStore (both to be discussed in detail later) are very relevant tools and have been applied time and again in both an open-ended format and a target-specific format. Open-ended means that all technology and know-how within a given company are measured towards all potential business directions, products, and services. Target-specific simply restricts the number of business directions, but still keeps all potential technical solutions to be applied towards the targeted business directions. Again, I shall discuss this in much detail in later sections.

This sector is also the area in which there is a healthy balance to be observed between consumer pull ("I know the goal") and technology push ("I know the technology") and it is like in a tug of war: it can go to either side. This balance between consumer pull and technology push was already mentioned in earlier chapters, for instance Chapter 5. I strongly emphasize this balance and the need to fully understand its importance in the development and business environments. It has to be recognized that there are always people behind both the consumer pull and the technology push, and I mean people inside your company, people who represent interests, more often than not very legitimate, well-founded, and grounded ones. I mention this here because it is not always easy to distinguish between very personal reasons and very logical ones in the best interests of the business. Even within one company you will have different divisions or businesses that emphasize one of the two over the other, and this can and does change over time as the business develops. There is no one answer as to what is the right balance between consumer pull and technology push in the food industry; you almost have to play it by ear and support what you believe in most and where you, an individual within a larger organization, can contribute best. Do not become frustrated when it doesn't go your way. It's almost like the weather in England: it changes every so often!

However, when you fight for your way, whether it is consumer pull or technology push, be well prepared with relevant competences, arguments, and vision, and do recognize, at any given moment, in which of the four segments of the innovation space you are acting. By doing so, you can ensure that you are anticipating the trend in your direct environment rather than running after it. This is part of the street-smartness that you might have had, improved, or acquired by reading thus far.

Figure 10.3 is an artist's representation of the balance between consumer pull and technology push, which ideally could be a perfect balance but most often is definitely tilting to one side or the other.

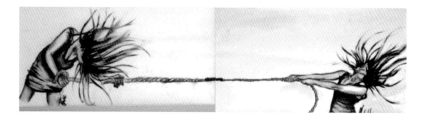

Figure 10.3 Consumer pull—technology push. Based on paintings from Meliné Katchi, Los Angeles, 2009. (Reproduced with kind permission of Meliné Katchi).

Quest

This SE quadrant describes the area in which the business knows the goal, knows the consumer target, and in which marketing has clearly identified, promising, and valuable new consumer offerings for which there is no technical solution yet. The important word here is yet, and the technical community, together with marketing goes on a quest to find the missing technological link. By definition, such technical know-how does not exist yet within the company (although as a technical person working in a food company you may have the feeling that you know it all and have it all …) and therefore has to either be created within or scouted from without.

This is the typical territory for long-term projects, involving multi-functional teams attempting to find the missing innovative and efficient technical solutions so that marketing and business people can realize their important goals. Immediately, one can see the problem arising: the timelines are completely out of sync. Whilst a long-term solution for marketing may last six to nine months, the dimensions of long term for technical development may be 24 months or, for greater complexity, 36 or 48 months. This latter timeline does not even take sometimes necessary clinical studies conducted in nutrition research into account; in that case we may even speak about a timeline of beyond five years. This is not an easy situation and is the most frequent reason why marketing and R&D in a food company fall out of love: time and expected timelines are very relative.

The main question here is: how can this be solved and how can the different expectations and perceptions of time be better harmonized so that both sides have a clear understanding of the situation at any moment? It's almost a million dollar question and one can get many different answers. Here is a selection of the ones heard more often:

- Develop faster
- Do not sell your project too early
- Do not over-sell your project
- Run projects with intermediate, usable results
- If it already takes a long time, be at least on time; plan better
- Get more resources, ideally without the extra costs
- Have better project management and project management skills
- Converge your efforts, work on fewer projects at any given time
- Get a better understanding of the consumer and in turn of marketing's needs
- R&D has let us down, again.

You may have a personal response list but I assume that it is probably similar to my list above. Unfortunately, there is some truth to all of these answers, as hard as it may sound, and I can hardly argue against them, other than to say: business people have to learn that innovation and innovative solutions to technical problems cannot be commandeered, execution is based on solid and persistent legwork, which requires skills, focus, and endurance, and often a good portion of luck. Having said this, I should immediately add that this should not be taken as an excuse for lengthy project deliveries and unfinished projects.

It is, however, very important to understand that both parties, business and R&D, have to find common ground and come together in the spirit of collaboration and with the clear aim to succeed in their joint efforts. This sounds like a pep talk, and maybe it is. On the other hand, it's the only viable and successful solution to this extremely important issue. Innovation tools such as FastPack or IdeaStore are ideally designed to help finding this common ground, not necessarily by accelerating the creative and innovative process, but by generating a joint understanding of the complexity of any conceptual development work very early on. Both parties thereby become "companions in crime" and have a much better understanding of what it takes to get this done, and jointly get it done until final, successful project execution.

Whilst working in the NE quadrant "Execution" is mostly about finding and fighting for the most appropriate balance between consumer pull and technology push, the main requirements in the SE quadrant "Quest" are creating understanding of the necessary timelines in complex development projects, in the quest for the right technological solutions and, moreover, involving business as well as development representatives early on in the process.

Movie

This NW quadrant of the innovation map is the typical territory for technology push. R&D people love it! They can showcase what they know how to do best and can impress the business side of the company. At least, that's what many in the development community believe. Stop: it's a trap! One can so easily get carried away with one's preferred pet science and technology that the most important question is not asked often enough: "How do we make money with this?" There are other, well-known questions that need to be asked in this situation, such as: "What's the size of the prize?" However, the latter already assumes that we can make money and is therefore a logical consequence of the first question. It is really *the* question and whatever good idea you might have, based on your preferred science or engineering know-how, always ask this question: How will we make money with this?

It is to be expected that you may not have an answer to these questions but by simply asking them, you will have to dig in deep in order to find the right answers in terms of product or service offerings to the consumers, operators, and/or the trade. It's not an easy journey and can be compared to the situation in the movie industry: they have the technical means and creativity to make movies, good movies, and even great movies; but a movie without content, without a great story, is no movie at all, but just a sequence of images, not enough to attract viewers. That's why the movie industry is always chasing after good scripts. Remember, when the Hollywood Writers' Guild went on a rather lengthy strike a few years ago, the industry almost came to a standstill. The strike probably took so long as there were still good scripts in the drawers of the producers and movie directors; once they had run out of these, they had to give in.

What we can do in the food industry is to be extremely well prepared, with as much consumer and business data as possible, paired with great technical skills, creativity, and innovativeness, and, last but not least, the guts to jump forward and create wonderful new products or services for the food industry. The "guts" part is often very much neglected and too much of a Cartesian, brainy approach is applied. The consumers stand on the other side of a ditch and there are two possibilities to reach them: either build a bridge by generating lots of consumer data and conduct consumer research or, based on your own confidence, you jump. The latter is not necessarily more risky than the former; both still leave you with lots of uncertainties. My personal recommendation, based on many years of observation and personal experience is to jump; you may want to build a small safety net though.

Fog

I kept the best for last. It's probably the most fun area and may also come out with the greatest rewards. Just imagine, you go to your boss and you tell her that you want to work on a concept idea that does not have any clearly identified technical solutions going with it yet and you do not even know for which target consumer or customer this would be developed. I would like to be there when you say this; I really would love to see your boss' face! It's not going to be an easy situation and not only needs a lot of guts on your part, but also a really good plan as to how you would approach this, and a number of convincing arguments, as well as allies to help you move forward in a meaningful way. Such allies can be colleagues who believe in what you are trying to achieve, namely going about ideation in a doubly open-ended way (open-ended on the target and open-ended on technical solutions). This approach can be especially appropriate when the company has a rather empty front end of the innovation pipeline and not enough other, more traditional and straightforward means to fill it. This might be the case, when a company is going through difficult times, has not enough success in the marketplace and/or may have gone through a series of reorganizations.

The most prominent example for this was P&G, when they invented and launched the Swiffer brand in 1999. This often-cited example of the invention of Swiffer came out of such a phase "0" type of project in which every search and every direction is permitted at the onset of the creative problem-solving process. P&G had the good foresight to jump on a technology from the Kao Company in Japan, which was not patented for this type of application, and which presented a unique and subsequently successful technology to achieve their target (Business Courier, 1999). In later years, they used IDEO to expand the Swiffer offerings. In this context IDEO wrote the following short story:

> ### Resetting the design DNA of the Swiffer Sweeper®
>
> Procter & Gamble came to IDEO for help designing a next generation Swiffer Sweeper®. The new tool was planned to build on the success of its popular forerunner with enhanced features and renewed appeal. To maintain its competitive pricing, the design was to maintain the basic Swiffer sweeper concept: a reusable tool with refillable cleaning pads. At the same time, the new sweeper design represented an opportunity to define the design language of future Swiffer® products for a stronger brand image, and a more consistent and delightful experience.
>
> Based on in-home observations, the multifunctional IDEO and P&G-team began designing around a number of opportunities for improvements to storage capability, handle comfort, cleaning ability, and appearance.
>
> The final design boasts a series of clever features, including a broader head for increased cleaning surface, and a reinforced, flexible hanging ring that facilitates storage and allows for slip-free leaning. The pole of the sweeper is larger in diameter and extends inside the handle for visibly stronger construction. The handle itself is more ergonomic to fit a wide range of hands.
>
> A fresh look at materials has given the design a layer of soft, translucent elastomeric material on the handle to provide a secure and comfortable grip for more control. A layer of flexible foam padding lining the head provides support even when the sweeper head is flipped for scrubbing action.

The overall aesthetic is one of flowing lines, with a dynamic form that alludes to the active nature of the product.

Project date: 2006

(http://www.ideo.com/work/swiffer-sweeper)

The Swiffer® line of products, with all its variations, quickly became one of the most successful brands in the consumer goods industry's recent history, and is performing well more than 10 years after its introduction into the marketplace.

So, we have gone full circle, we did a "360" and gave the innovation space a closer look. As I have already stressed in the introduction to this section, it is extremely important for anyone who intends to propose conceptual work towards new product or service offerings to know into which of these four quadrants the proposal would fit and act accordingly, when it comes to applying the most appropriate vehicles in terms of required expert resources, as well as innovation tools that can help to speed up the process and thereby get into the marketplace much quicker.

I would, however, stress that using appropriate innovation tools and supporting methods in any of the four quadrants can substantially improve the chances of getting the job done and push your innovative concept successfully through your organization to the finishing line of execution and launch. The choice of which tool or method you will ultimately apply is more often dependent on your budget and the type of allies that you find in your company, and who, like you, want to speed things up and make the entire chain of events in the process more effective.

FROM BRAINSTORMING TO CREATIVE PROBLEM SOLVING (CPS)

Many of you must have experienced brainstorming exercises at one time or another during your professional life. I do remember those, where we came up with hundreds of ideas, many of them probably even very good ideas, and we were very proud of our achievements. What achievements? Well, the number of ideas, look at them, and someone should probably do something with them, and we look forward to great successes, and let me know, in case I can be of any further assistance … Sounds familiar? Probably does, as this is, or at least was, most likely the most prominent scenario and outcome of such brainstorming sessions: pride for the number of ideas, but no real follow-up. Brainstorming is an old and proven technique, however it needs to be carried out in the proper way.

Arthur B. VanGundy in his important book: *Idea Power: Techniques and Resources to Unleash the Creativity in Your Organization*, developed a manual for CPS, in which brainstorming plays a central role; structured brainstorming, that is (VanGundy, 1992). What do I mean by structured brainstorming? In order to answer this question, let me elaborate a bit more on CPS, as defined by VanGundy. He distinguishes between the act of creative problem-solving and the methods that derive from it, such as the CPS tools developed by VanGundy himself, as well as Alex Osborn and Sidney Parnes. Osborn, also often called "the father of brainstorming," wrote the landmark book on CPS, *Applied Imagination* (Osborn, 1953/2001). Parnes subsequently crafted an organized, six-step process for CPS. These steps are: Objective Finding, Fact or Data Finding, Problem Finding, Idea Finding,

Solution Finding, Acceptance Finding. I will discuss these steps in more detail below and especially how we developed and adapted the process so that it could and can best fit the requirements of a large food company. The six steps can be organized into three groups:

- *Explore the Challenge*: Objective Finding, Fact Finding, Problem Finding
- *Generate Ideas*: Idea Finding (the actual brainstorming)
- *Prepare for Action*: Solution Finding, Acceptance Finding.

(Creative Education Foundation, 2013)

There are more creative processes and accompanying methods and tools to be found, such as at Stanford Research International (SRI) in Menlo Park, California, led by the very determined Curtis Carlson, CEO and President of the Institute. For detailed reading, consult his book *Innovation: The Five Disciplines for Creating What Customers Want* (Carlson and Wilmot, 2006). To give you some insight into SRI's thinking, here's a short quote from the VP and General Manager, Stephen Ciesinski:

> *Innovation is one of those words greatly in danger of overuse and exaggeration. These days, you hear the term mentioned in almost every corner of industry and even in reference to culture, society, and our personal lives.*
>
> *At SRI, we don't use the word "innovation" lightly. We have a clear and crisp definition for what it means*: the creation and delivery of new customer value into a marketplace, with a sustainable business model for the organization producing it.
>
> *My organization at SRI is chartered with commercial and international business development. We have the unique perspective of seeing innovation unfold at the global level when we travel to places such as Japan, Chile, the Mideast, Turkey, the Nordics, and South Korea, as well as throughout the USA.*
>
> *We partner with the world's best and brightest innovators, researchers, and entrepreneurs to share SRI's best practices for innovation process—a process we've found central to the growth of our company.*

(SRI: http://www.sri.com/blog/sris-unique-innovation-charter#sthash.0afSpC1E.dpuf)

The major direction that SRI takes is to involve as many stakeholders as possible in the creative process, then go through several iterations and make sure that only the most valuable ideas are pursued and brought to a successful end. Curt Carlson, during a personal meeting, specially mentioned their mantra to only go after the really big opportunities, even targeting as high as the billion dollar mark (C. R. Carlson, personal communication, 2008).

THE DIVERGENCE–CONVERGENCE PAIN

Brainstorming is all about letting your creative juices flow and loosen up as much as you can, and some more. This is the divergence part of the entire exercise. You should go out on a limb and there should be no limitations and especially no negativity with regards to the ideas that you and your colleagues come up with during this process. Once you have converged enough or, in other words, have collected a seriously substantial number of potentially good and meaningful ideas, you have to start the convergence process, i.e. eliminating the less meaningful, less relevant, and potentially less rewarding ideas and only select the very few that meet the set of criteria that you have defined beforehand.

This set of criteria covers several steps of the traditional brainstorming process: objective finding, data finding, and problem finding. The brainstorming itself is the act of idea finding and, based very much on personal experience with brainstorming exercises, it is crucial, especially during the very open-ended process of idea finding, to always remind participants of brainstorming exercises of the relevance of the jointly developed problem finding results. Typically, organizers of a brainstorming exercise set out a set of objectives, or in other words, what is it that we are jointly after? Typically this set of objectives is called the "Preliminary Wish Statement" and is formulated in a specific way: "I wish I could achieve … " or, more recently, we have used the formulation "It would be great if we could … "

With input from everyone involved and special guidance from the brainstorming session organizers, the preliminary wish statement is then refined, based on relevant and pertinent data, and taking the competences of the different participants into account, emphasis is put upon coming up with ideas during the brainstorming that are relevant to what the team can actually tackle and resolve. Based on the learning from many brainstorming sessions, we have translated the problem statement into a "Final Wish Statement" in order to show the continuity from objective setting and preliminary wish statement to data input, and to the final wish statement. In actual fact, any brainstorming session starts with a divergence—convergence exercise, i.e. first diverge through the preliminary wish statement and then converge by adding meaningful data and narrowing it down to the final wish statement. The group, and brainstorming is typically done in groups, then sets out on the true divergence exercise itself, the brainstorming, however always keeping in mind the relevance and statements spelled out in the final wish statement. The leader of the session has to make sure, without any negativity in mind, to guide the participants in the right direction. This typically can be done in amusing and playful ways; there are many ways to happiness, to a good final result and substantial quantitative, as well as qualitative output of a brainstorming exercise.

Once the idea finding part is done, participants typically feel drained and have a sense of pride and achievement. However, the important part really only comes now, the pain of diverging and converging again and selecting a small number of really good, promising, relevant, and potentially rewarding ideas. This is not an easy task and there is no one way as to how to achieve this. Some people have used some kind of "objective metrics," applied by committees and other intelligent groups. From personal experience, I can say that the most efficient way to do this is to apply some democracy, maybe one of the rare moments during which you can find democracy in a large company and it's working teams. Based on results from many brainstorming sessions that I have personally participated and/or led, the best way to go about this is to organize these sessions over a period of two days: reserve the first day for the preliminaries and the idea generation. The participants, pretty substantially worn out after such a day can then sleep on the events of the first day and proceed to a personal voting process, by which the majority of votes given to the various ideas brings out a winner and runners-up, etc. Typically there is some natural clustering, and it is common that out of a number of, for instance, 300 ideas generated during the first day, there are only four to eight really relevant and promising concepts that have been selected by the participants and that are worthwhile to pursue in more detail.

This selection process is probably the most painful exercise as it often means saying goodbye to some ideas of my own, to pet ideas, or to something that I wanted to do for a long time and believed that this brainstorming session may have been the occasion to get it approved. From experience, it is wise to not come to such an exercise with too many, if any, preconceived ideas, but really let your creative juices flow untainted, except for your personal experience and the relevance to the final wish statement. Once this not so easy convergence part is done, one typically moves on to the next step, which is the part of solution finding; we call it elaboration of the concept through group work. As time is a limiting factor, every group that works on more than one idea, and that is typically the case, is confronted with this time constraint and has to work under pressure. I will discuss this in more detail later on, when I talk about the IdeaStore.

The very last step of every brainstorming is selling the selected and elaborated idea, the acceptance finding. It's like every sales job, you have to be well prepared, have your facts together, be concise, and be convincing. The latter requires the famous "shine in the eye." Without it, you may be best prepared, but will still not get your great concept idea accepted by the decision-makers, typically your boss, your marketing or sales colleague, your procurement colleague or a mix of these. This is not to say that with the best "shine in the eye" you don't need to prepare well, not at all, it requires both: best preparation, obviously based on a great concept idea together with your shine, your passion for it. You clearly have to show the decision-makers that you really believe in the concept and that it's going to be a great success. Getting acceptance is not an easy task and often leads to quite some frustration because you were rejected. There is no one recipe to escape this rejection—frustration cycle, the best one is probably to never give up and try harder next time.

FASTPACK: A BRAINSTORMING EXERCISE SPECIALLY DESIGNED FOR PACKAGING DEVELOPMENT

FastPack was developed by three people in the Nestlé company, with help from the design agency M-Design of Burbank (formerly North Hollywood), back in 1999. It is fully based on Osborn's brainstorming principles, but perfectly adapted to the requirements of efficient and successful packaging development. All three Nestlé USA colleagues, Mike Rodero, Bob Lepior, and Gordon Lane worked in packaging and were frustrated with the way packaging development was done in those days, namely by numerous iterations and frequent rejections by marketing and sales colleagues. This led to extreme inefficiencies and especially costly time delays. Very early on they recognized the need to get a designer on board and so it happened that M-Design's Spencer Mackey became an important fourth member of the team that created *FastPack*.

In this particular variant of the problem-solving method originally developed by Osborn, there are two extremely important additional features that make *FastPack* a unique and a uniquely successful innovation tool. First, it is part of the genes of *FastPack* to have as diverse a group of participants at the event as possible. Secondly, the event is led or at least co-led by designers. The designers, typically two that complement each other, are crucial to the speed of the entire event, which consists of profound preparation beforehand, the *FastPack* event itself, and the work that comes afterwards and includes in fast succession: writing a concise yet meaningful report, collecting all relevant event data, such as notes and drawings, proceeding to fast prototyping of the top three or four selected packaging ideas, developing concept boards, and, together with marketing colleagues, performing preliminary consumer research with the material at hand.

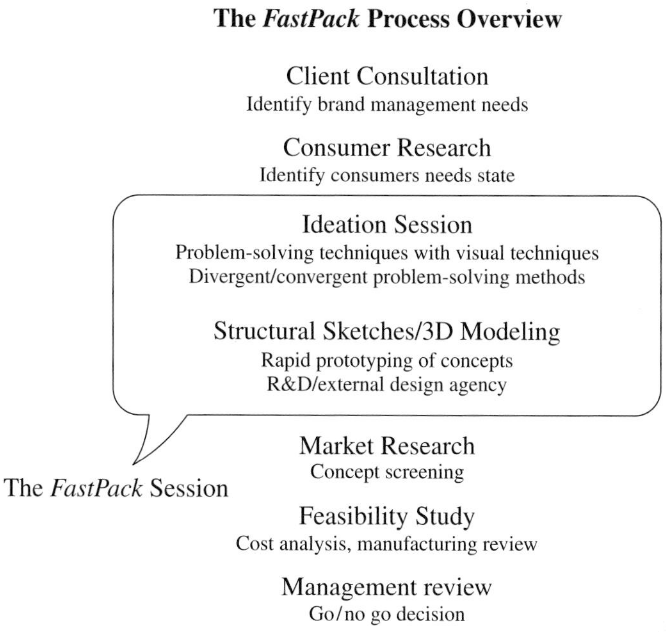

The *FastPack* Process Overview

Client Consultation
Identify brand management needs

Consumer Research
Identify consumers needs state

Ideation Session
Problem-solving techniques with visual techniques
Divergent/convergent problem-solving methods

Structural Sketches/3D Modeling
Rapid prototyping of concepts
R&D/external design agency

The *FastPack* Session

Market Research
Concept screening

Feasibility Study
Cost analysis, manufacturing review

Management review
Go/no go decision

Figure 10.4   The *FastPack* process—sequence of actions—overview.

One additional feature that was introduced to *FastPack* at a later stage is to give each participant the challenge of "homework," to engage participants mentally as well as possibly physically before they arrive at the *FastPack* session. By physically, I mean that each participant is asked to prepare one good idea, one good example of a relevant packaging concept, either in the format of a sketch, or as a hand-crafted small model, as a starting point for the workshop. The entire process can be done in really fast mode, i.e. typically only six weeks for all the steps described above, ready for the final decision by the marketing colleagues. You can see that the name *FastPack* was not chosen lightly, but really reflects the efficiency of this great innovation tool, specially adapted for fast and successful packaging development. Since its inception it has probably been used approx. 200 times and numerous packages were developed based on the outcome of *FastPack* sessions. Figures 10.4 and 10.5 show a short overview over the *FastPack* process.

THE *IDEASTORE*

The *IdeaStore*, in its original "generation one" open-ended format, is yet another example of a bespoke innovation tool, perfectly adapted to concept, project, and ultimately product and process generation in the food industry. I was one of the co-creators, together with Spencer Mackay from M-Design and the *IdeaStore* is crafted in the mold of *FastPack*. The main differences are:

- Contrary to *FastPack*, where concept selection is more severe, down to only three or four concepts, *IdeaStore* allows for six to eight concepts. This is simply because of its open-ended format, and although there is strong business data input from the outset,

The *FastPack* Process Attributes

Speed
• Rapidly identify and screen opportunities
• Development time rapidly reduced
• Rapid execution of 3D CAD modeling

Confidentiality
• Process company owned and driven
• Managed exposure to outside agencies
• Process identifies proprietary opportunities

Cost saving
• Controlled external agency costs, by leveraging in-house expertise
• Cost to develop concepts to SBU / business decision minimal

Teamwork
• Strong partnership with R&D & external agencies
• Process brings marketing, sales, manufacturing, engineering,
  purchasing and R&D together from initiation to implementation

Global Flexibility
• *FastPack* is a "tool" easily transferable to many different company locations
• It is a process that can work for packaging, product, or process innovation

Figure 10.5 The *FastPack* process—attributes—characteristics and potential.

specific relevant businesses or business units are brought on board only on the basis of the selected results of the *IdeaStore* (generation one). Attrition by the businesses is typically rather severe: from experience, two to three concepts make it to projects, out of which one or two make it to the finish line of products or processes.

• Brainstorming takes place in defined product, brand, process, and service areas; it is not packaging driven.
• The selection by the businesses takes longer than in the case of *FastPack*, and, based on experience, may take approx. two to three months before it is agreed to move from concept presentation to starting a project.

Figure 10.6 illustrates a typical flow chart of the *IdeaStore* along the time axis: preparation before the event, two days of actual *IdeaStore*, and follow-up actions.

In an article "Creating Successful Innovation Partnerships," on pp. 32 and 33, together with my co-author Sam Saguy, I wrote the following about the *IdeaStore*:

One creative-problem-solving tool is IdeaStore. *It is a variant of brainstorming. It is based on three main characteristics*:

1. *Facilitation by "visualizers",*
2. *Large diversity of participants, and*
3. *Disciplined and stringent selection and prioritization process.*

IdeaStore *uses well-known brainstorming techniques, combined with clear goal setting, diversity, divergent and convergent idea flows, stringent and fact-based selection, as well as the thorough elaboration of concepts followed by prioritizing the selected ideas. Its main thrust is*

The Ideastore

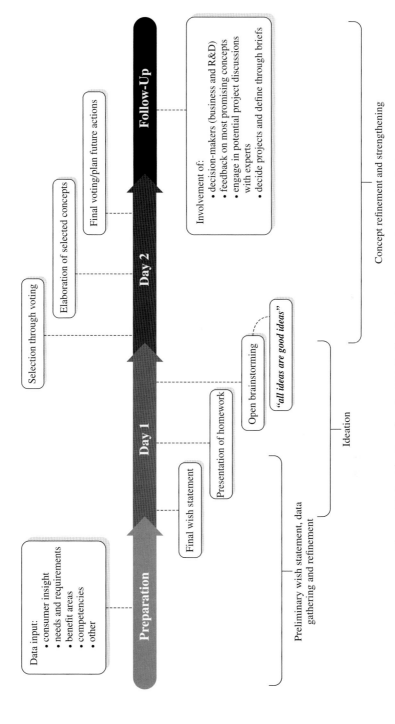

Figure 10.6 *IdeaStore* flowchart (H. Traitler, S. Mackay, personal communication).

to discover or to create opportunities by matching needs and requirements with relevant competences. An IdeaStore *session typically takes place over 2 days. The session brings together an approximate equal number of representatives of both partners.*

[*Note from the author*: this is valid in the case of *IdeaStores* carried out between your company and an innovation partner. I shall discuss the concept of Innovation Partnerships in more detail in Chapter 11. *IdeaStores* can be run within your own company; however, it is crucial to always respect as much diversity of participants as possible.]

For practical reasons, the workshop is led by two facilitators who have a design background as they also act as visualizers. They can capture and visualize images that help better explain the potential concept and give a possible project a face.

The total number of participants should typically not exceed 14. This number enables every participant to creatively contribute to the brainstorming, selection, elaboration, and prioritization process—crucial elements of the IdeaStore *exercise. The background of the participants of the* IdeaStore *session should be as diverse as possible. Based on experience and the business focus of Nestlé, a nutrition savvy person is always included. Furthermore, Nestlé always involves a person from procurement so that from the beginning, commercial thinking can flow into the ideation process. Finally, people who can make decisions when the time comes to dispatch necessary resources into projects are also involved.*

Note from the author: We always involved a representative of the Nestlé procurement team, by this assuring that those who ultimately had to make purchase decisions were involved very early on and demonstrating the complexity, or simplicity, of the expected development to them.

Every IdeaStore *process begins with the definition of the preliminary wish statement. The wish statement typically might read: "It would be great if we could … "*

The statement will comprise short-term as well as long-term goals. When formulating the statement, organizations must consider such factors as value of concept (the aforementioned "size of the prize"), whether the concept covers a sufficiently big business opportunity, availability of resources, realistic timelines, consumer benefits, etc. The wish statement might also deal with "soft targets" such as building the team spirit between the different participants.

In order to improve the chances of successful follow-up of any of the concepts developed during the IdeaStore *exercise, it is advisable to limit the number of concepts in the selection process to not more than 10, based on experience rather 6 to 8. Once such potential concepts have been elaborated and brought into the format of a concept sheet, they are submitted to the relevant businesses and R&D groups for validation of relevance. The typical success rate of validation is around 30% (i.e., approximately 2 to 3 projects are being further elaborated and jointly worked on between the various stakeholders). The pattern that emerges from such workshops can be described typically as: 250 to 300 → 6 to10 → 2 to 3 ultimately selected and worked on projects.*

IdeaStore *workshops also create an important, almost intangible by-product, namely team building and an open invitation to continue along this spirit of innovative partnership beyond the session. Once the groups interact, they start to know each other better, they understand what each group can bring to the relationship, and the interaction opens new pathways to direct co-creation."*

(Traitler and Saguy, 2009)

Like in the case of *FastPack*, we introduced the quest for a homework idea to be concocted and brought to the *IdeaStore* as a starting point for the brainstorming session. From experience, and when *IdeaStores* were jointly run between the company and one of its partners, each *IdeaStore* brought the above- mentioned two to three important, relevant, and agreed upon projects as a direct outcome of the session, but it additionally created a similar number of joint projects, simply based on the fact that the participants got to know each other and started to interact directly, and thereby discover other important business opportunities. The Nestlé company carried out approximately 20 *IdeaStores* between 2008 and 2013; 16 were run in open-ended (generation one, gen1) mode, i.e. all possible competences represented, potentially for all businesses and for all geographies (markets). Two generation two (gen2) *IdeaStores* were carried out with a focus on two, maximum three, businesses and, for instance, the geography of North America (mainly USA and Canada) with participants with relevant backgrounds. A further two gen1 *IdeaStores* were performed in 2013 within the boundaries of the company's R&D capabilities, very much technology pushed, but with important and substantial consumer insight flowing into the preparation and the establishment of the preliminary wish statement.

As far as the 16 gen1 *IdeaStores* are concerned, a kind of steady state total number of active projects was very quickly generated and leveled off at around 70 to 80 projects two years after the first *IdeaStore*. You have to bear in mind that this number was the result of the projects coming directly from the *IdeaStore* together with the others that were created on the basis of the newly created relationship between the participants. I shall discuss generated resources as well as derived business in the next chapter, "From open innovation to partnerships."

INSIDE THE BOX

For whatever reasons, there seem to be voices out there in the industry that challenge brainstorming and suggest that people are most innovative when they work within the boundaries and constraints of their own, existing knowledge as the foundations of their creativity and innovativeness. Drew Boyd and Jacob Goldenberg published this theory in their book *Inside the Box* (Boyd and Goldenberg, 2013). The *Wall Street Journal* wrote an almost two-pager on this book in June 2013. The article starts with the following paragraph: "When most CEOs hear the word "innovation", they roll their eyes. It conjures up images of employees wasting hours, even days, sitting in beanbag chairs, tossing Frisbees and regurgitating ideas they had already considered. "Brainstorming" has become a byword for tedium and frustration". This rather derogatory statement makes for a catchy introduction to a book that actually talks about something else, namely the field of tension between individual contributions versus team contributions when it comes to creating innovative solutions. Almost more importantly it talks about a series of techniques that can make innovation potentially more successful. In the second paragraph of the *Wall Street Journal* article, you immediately find a contradiction to the first paragraph, namely that for all (or most) CEOs, innovation is rated very highly, however when it comes to innovating, there is still much to be done. Bingo, I totally agree with this. However, then it goes on by saying that "the traditional view of creativity is that it is unstructured and doesn't follow rules or patterns," (The Wall Street Journal, 2013).

In the earlier sections of this chapter I laid out and discussed the tediously crafted ways of how modern innovation can work through highly structured, almost ritualized tools. These are all but "by chance" tools, they follow almost religiously the structured pathways that have led to many innovation successes in the past and still to this day for many companies, especially food companies. I can understand that some CEOs "roll their eyes", because if someone sells them the theory that every individual in every company can, on the basis of his or her knowledge base, come up with great innovations and that this is at no extra cost, "innovation at zero costs," they will immediately tell you: "That's great, no extra costs, I like that." However, really smart CEOs would not even get involved in such a discussion, as they might have more important activities to pursue than to decide whether brainstorming or "inside the box," in solitude, is the better way to successful innovation.

So, let me get to a few of the basic principles that Boyd and Goldenberg write about in their book. They propose that thinking "inside the box" brings people to really focus on the essential, recognized limitations and boundaries, and by this, they can become more creative. They use the well known "box metaphor" for this and consistently ask the question of where one has to be (inside or outside) at any given time in order to be the most creative and innovative. I have discussed the value and importance of constraints in Chapter 8. I totally agree with the strength of the argument that when boundaries within which I can work are constrained (in meaningful and realistic ways), I typically come up with a better, more adapted, and probably simpler solution. This, however, is by no means an argument to say that therefore, people should not think and work in teams anymore, and should innovate and create in the splendid isolation of their own minds. It appears in reading this book that simple and readily available solutions to problems are rejected because they are outside the box and we are encouraged to pursue more complex solutions, which are to be found within the constraints of only inside the box, and all this just to make the point.

I personally believe that the combination of constraints (as for instance expressed in a concise and precise wish statement of an *IdeaStore* session) the creative, structured, and guided combination of minds, and the multiplicity and diversity of know-how, is an almost guaranteed path to success when it comes to finding the next great innovation for your company. Boyd and Goldenberg propose a specific method for "Inside the Box" that is based on five principles or techniques:

- *Subtraction* → Remove seemingly essential elements
- *Unification* → Bring unrelated tasks or functions together
- *Multiplication* → Copy a component and then alter it
- *Division* → Separate the components of a product or service and re-arrange it
- *Dependency* → Make attributes of a product change in response to changes in another attribute or in the surrounding environment.

I believe that they are on to something as far as these steps are concerned, and I would suggest that you dig deeper into these five principles and apply them in the most appropriate ways in your environment. However, these principles do not depend on whether you do this in solitude ("Inside the Box") or as a team in a highly structured brainstorming exercise, such as *FastPack* or *IdeaStore*. I take the pragmatic approach here: whatever works best for you, is probably best for you. You can try out the solitude approach and when I discussed the precursor rituals for innovation tools earlier in this chapter, many of these rituals are performed in solitude, even under very constrained circumstances, especially when it comes to the spatial rituals. There is no contradiction at all between doing this as well as the structured, well-guided team approach of brainstorming: you probably already do this to some degree every day at home, as well as at work.

CONCLUSIONS

In this chapter we discussed and analyzed all sorts of innovation rituals and tools, we discussed and analyzed ways in which the innovation space could and should be organized, we discussed and analyzed in detail some of the origins of creative problem-solving and brainstorming, the principles of divergent and convergent thinking, and two specific and well-tested examples of creative problem-solving, namely *FastPack* and the *IdeaStore*. We can conclude the following:

- The first step in structured innovation tools or methods are all the well-known and maybe less well-known rituals, sometimes even personal "ticks," which I have grouped into: spatial (e.g. shower)—irrational (e.g. wear a hat)—situational (e.g. cook a meal). I suggested that the reader finds out those very personal rituals that work best for him or her and applies these in consistent ways, almost like an athlete who trains every day.
- It is important to understand the innovation space. I used an example from the P&G company that mapped its innovation space into four quadrants and suggests approaching innovation in different ways, those most adapted to the specific locations in the innovation space. The four quadrants are defined by the two axes: knowing the goal (consumer) versus not knowing the goal, and knowing the means (the technology) versus not knowing the means. The four quadrants are named after the most logical approach, i.e. "Execution" for knowing both the consumer and the technology, "Quest" for knowing the consumer, but not having the technology at hand, "Movie" for knowing the technology, but not yet knowing the consumer, and finally "Fog" for the situation in which we neither know the consumer (yet!) nor the technology (yet!).
- I briefly introduced, discussed, and analyzed some of the origins, as well as principles of creative problem-solving, in which traditional brainstorming plays a central role.
- I discussed and analyzed the importance of a succession of divergent and convergent thinking in an almost painful manner when it comes to proper brainstorming. I emphasized the crucial importance of selection of concepts measured against the initial goals and relevance to the business. I discussed step-by-step how a typical, successful problem solving event should be structured.
- I introduced and discussed the example of *FastPack*, a well-established creative problem-solving tool, invented by a few people at the Nestlé company, mainly out of frustration with the existing situation of "ping-pong" between marketing and technical packaging development. *FastPack* was invented around 1999 and has been used approximately 200 times since. The role and importance of design and designers leading such an event was clearly emphasized.
- Crafted on the experience and the structure of *FastPack*, the *IdeaStore* was created in 2007. *IdeaStore* uses the same principles as *FastPack*, with the major difference that it is not packaging solution-driven but works in the innovation space of products, processes, and services. The *IdeaStore* is ideally led by designers, who play the role of

visualizers. The *IdeaStore* exists in two major formats, generation one and generation two. Gen1 is the open-ended version, gen2 is the business- and geography-centered variant. As in *FastPack*, the greatest possible diversity of participants makes for a greater success rate.

- I discussed and analyzed the principles of a recently (June 2013) published book *Inside the Box*, in which almost the counter-thesis of team-based brainstorming is suggested. It is postulated by the authors Boyd and Goldenberg that people can be most innovative within the boundaries and restrictions of their own, existing knowledge base, and may come up with solutions that are cheaper and more adapted to the needs of the company. Although I am critical of the solitary innovation process, the authors propose a set of techniques that can be useful in all types of innovation processes, including well-structured traditional problem-solving.

TOPICS FOR FURTHER DISCUSSION

- Discuss and critically analyze, in your personal environment, with family and friends, your own personal rituals during which you feel a boost in creativity and innovativeness. Where do YOU have your BEST ideas? Try to fit your rituals in the chart in Figure 10.1.

- Discuss and critically analyze the validity of the proposed mapping of the innovation space. Do you apply something similar in your work environment; does your company apply anything at all? How does your company go about the balance between technology-push and consumer-pull? Did you experience any accusations from one side or the other, depending in which area you work? Discuss with your colleagues what balance, if any, could be best for your company.

- Take the situation of "Fog" from the innovation map, and analyze this specific situation in more depth and discuss with your colleagues how your company could most profit from especially this type of location on the innovation map and all the missing information. Does your company have the courage to be active in this field?

- Discuss and analyze the situation of creative problem-solving in your work environment. How much of this is done in your company? If not enough, discuss ways to improve the situation and have a plan for increased success based on such techniques, including brainstorming and severe selection. Do not forget to discuss why you believe that such techniques may not be necessary, collect the arguments, and weigh them against the arguments for creative brainstorming. Measure the results.

- Discuss and analyze your personal experience, possibly "traumatization" and frustration, with successive divergent and convergent thinking. Do you already do it unconsciously? If yes, analyze the results that it may have brought. Compare, discuss, and analyze structured versus unstructured ways for the divergent-convergent thought process.

- Have you ever participated in something similar to the proposed innovation tools *FastPack* and/or *IdeaStore*? What was your experience? Critically analyze the situation in your workplace and suggest improvements to the actual situations. Or discuss the arguments why you and your company believe that it is not worthwhile pursuing. Be critical, with pros and cons!

- Analyze and discuss the suggestion to innovate within one's own knowledge base, in isolation. Discuss your personal experience with your colleagues and try to combine the two approaches of "isolated" versus "structured teams." If you find convergences or even overlaps, please let us know!

REFERENCES

D. Boyd, J. Goldenberg (2013). *Inside the Box: Why the Best Business Solutions are Right in Front of You*, Profile Books Ltd., London.

Business Courier (1999). New Swiffer cleans up for Procter. Bizjournals.com.

C.R. Carlson, W.W. Wilmot (2006). *Innovation: The Five Disciplines for Creating What Customers Want*, Crown Business, Crown Publishing Group.

Creative Education Foundation, 2013; http://www.creativeeducationfoundation.org/our-process/what-is-cps

A. Osborn (1953/2001). *Applied Imagination: Principles and Procedures of Creative Problem Solving*, Creative Education Foundation Press.

H. Traitler, S. Saguy (2009). Creating successful innovation partnerships, *Food Technology*, 03.09, www.ift.org.

A.B. VanGundy (1992). *Idea Power*, AMACOM, New York.

The Wall Street Journal (2013) June 15–16, 2013, pp. C1–C2.

11 From open innovation to partnerships

> *But think of the last guy. For one minute, think of the last guy. Nobody's got it worse than that*
> *guy. Nobody in the whole world. That guy … he's so alone in the world that he doesn't even*
> *have a street to lay in for a truck to run him over. He is out there with nothing.*
>
> Arlo Guthrie, The Pause of Mr. Claus, Appleseed Music Inc. (ASCAP), 1968, 1969

FROM OPEN INNOVATION TO PARTNERSHIPS: A LOGICAL TRANSITION

Open innovation quickly became the buzz-word in the early 2000s. Many books were written on the topic and P&G's "Connect and Develop" (C&D) was the talk of the town. Every company that I came across and I knew of at least talked about the possibility of doing something similar, and many companies pretty much crafted their own open innovation activities on C&D. Like most of the other food companies, Nestlé looked at the model very carefully, and a very thorough analysis was performed. The results of this analysis brought us, in those days (October 2006), to the conclusion that Nestlé would not ride on the wave of open innovation in a C&D mode, but would rather focus on creating solid and sustainable partnerships with key suppliers, academic institutions, and even individual innovators.

Before I go into this in more detail, let me discuss and analyze some of the main and important features of C&D, without wanting to repeat what has been amply published and discussed about C&D in the past, especially in Lafley and Charan's book, *The Game Changer* (Lafley and Charan, 2008). This book and numerous publications have not been the only sources for our analysis and we dug pretty deep during very friendly meetings and discussions with our colleagues at P&G. Back then (2005 to 2006) P&G was not considered a competitor, except for the area of pet-care, so it was unproblematic to openly discuss and share experience, provided we did not touch upon anything in pet-care.

Here, in short, the story told by a former P&G insider, who was involved in the creation of C&D:

> *What we discovered and found out was rather the obvious. The way in which C&D was*
> *designed required resources and time, and when I say resources, I mean really quite a*

Food Industry Design, Technology and Innovation, First Edition.
Helmut Traitler, Birgit Coleman and Karen Hofmann.
© 2015 John Wiley & Sons, Inc. Published 2015 by John Wiley & Sons, Inc.

substantial amount of these. We learned that in those days P&G employed a rough number of 130 employees directly as "technology scouts" (approx. 70) and the others in back offices in supporting functions (deal makers, compliance specialists, lawyers, admin, etc.). It appears that P&G's publicly known C&D program developed from two distinct groups: (1) technology scouts in the R&D organization who developed and maintained connections with research organizations, suppliers, and potential development partners, and (2) a business development group. While the first group does exist in some shape or form in every meaningful R&D organization, the latter was a unique construct. This group was set up separately from the technical functions and reported into corporate finance with an initial task to create value from P&G's investment into the R&D program. At that time, P&G spent globally about $2 billion USD in 28 research centers on R&D, held about 36,000 global patents, and yet some of the developments never made it to market and the projects were stopped at different stages in the development and qualification process.

This small group was formed with experienced and senior staff from different functions (R&D, Sales, Marketing, Finance, and Supply Chain) and supported by a small licensing support team (legal, contract administration, external relations). In the beginning, the primary task was to gain value from intellectual property, which has been generated in the R&D processes but never made it to market. Surely it must have been a cultural change to make patents and know-how available outside the own organization and this could have only worked by flowing back the financial benefits to the business unit, which owned the IP. So it was an enabling structural element to have this group report into the corporate finance organization as then it was easier to engage in concrete conversations about the use of the IP—either it was required for the business plans or it was 'fair game' for value generation—no more 'keeping just in case'. As money exchanged hands and this group created some form of income for the business without significant operational effort, it was gaining more attention from the business leaders and more attraction within the normal structure.

From there on, this group developed new approaches and business models. While the primary task was to access external innovation for use in the own product portfolio, it generated all sorts of deals, e.g. i) out-licensing trademarks to adjacent product categories in which P&G would never invest (competitive advantage), ii) combined IP/trademark deals to bring products to market under P&G brands which were too small for the big P&G organization but still meaningful for business partners (making best use of the existing assets), iii) qualification models which could help get a proposition to market such that the product was launched by a business partner with an option to spin it back into P&G under certain conditions .

This variety of tailor-made deal solutions apparently generated repeat business—at that time it seemed that about 40% of the business partners had multiple deals with P&G.

This activity, combined with the broad advertising of the C&D program made P&G an attractive partner, as it was apparent that there was money generated for P&G and the business partner.

It also created a very lean, experienced, and effective organization which was able to create partnership models quickly and get them aligned with top management rather than working its way through the corporate structure. Coming back to the core business, P&G explained at that time that important innovations like the Swiffer Duster®, Pringles Stix®, Olay Regenerist Derma-Pods®, and many more have their roots in other companies and have been accessed through deals which were enabled by this group.

So in net, this group seems to have been the transactional enabler for the C&D program and 'paid for itself' as it materialized value from otherwise mostly sunk cost and helped build a structure of open innovation and fruitful partnerships.

(K. Taubert, personal communication, August 2013)

Finally, an important feature of C&D, as for other similar approaches such as "Innocentive", "Innovation Exchange", "Ninesigma", "YourEncore" or "yet2.com," is to have one important leg anchored in the Internet. Outside experts who believe that they have some innovative and functioning solutions for the problems that the company has published on the Internet in a rather formalized way, can link in and connect with the company. All proposals are typically checked and valued and everyone receives an answer or at least an acknowledgement. From my experience, there are typically 4,000 such connections via the Internet per year, which ultimately leads to a rather time consuming validation and answering process.

All this explains the rather copious need for appropriate resources, hence the fairly high number. Most companies apply a number of approx. $300,000 for one FTE (full time equivalent), an overall amount that each individual employee costs the company. Unfortunately it's not the employee's salary, even if many of us wish it were the case. Anyhow, by multiplying 130 by 300,000 we come to a total apparent cost of $39 m per year just to pay all these resources. No risk, no fun, and apparently for P&G the fun was much bigger than the risk. Because this C&D group was not only in charge of discovering the next great and rewarding innovation out there, but also responsible for out-licensing non-core, non-critical know-how and technologies to the outside, we learnt that P&G made something in the order of $2 bn revenue over a rather short period of time. By looking at both sides now, one can better understand why the in-licensing part, the very trendy C&D, was much talked about as the greatest thing since the invention of sliced bread, whilst the very quiet out-licensing activities brought all—or most—of the money.

I am not aware of the exact amount of new revenue that P&G generated solely based on C&D per year—maybe I have not done enough research—but my feeling is that it's way below the $2 bn revenue from out-licensing. This leads me to conclude that an open-innovation approach such as C&D can only really work if it takes the two directions, in and out, into account, so that the necessary resources can be justified and the financial risk of failure is reduced substantially. It is pretty much the same as venture capital investment: the first step of investing in a start-up company is really (and I mean *really*) risky. This part is comparable to the in-licensing of know-how against invested money. The out-licensing can be compared to the exit of a start-up company: if all goes well, really good money can and will be made, much more than the investment. The real difference here is the following: The company (P&G in this case) knows the potential value of its unused know-how beforehand—it's almost like already knowing the most likely outcome of an exit at the time of investing—so, it's a much safer bet. This part is actually the one that I would highly recommend to companies, especially food companies: clean out your patent portfolio, look for the "re-imbursables" (a term used by Jet Propulsion Laboratory JPL in Pasadena for their out-licensing activities) and go about this big time. The rewards are probably much bigger than you think. The resources that you may need to do this are normally very well used and justified.

I spoke about two requirements: resources and time. I have not touched upon the time aspect yet, however it is extremely important and needs to be addressed. It takes time to do all this, I mean C&D and similar approaches and the successes are not immediate, if at all. For your management, time, based on personal experience, does not have the same dimension as for you; it appears to pass by much faster, and what is six months for you is half an eternity for your management: simply too long! And it is probably worse; it always takes too long, whatever your timeline may be. The time it takes to run a project and successfully implement it is probably one of the most controversial aspects of product and service development in the food industry. By default, it always takes too long. And I am strictly speaking about the time it takes for a project to run from end to end, i.e. consumer data gathering, ideation, conceptualization, technology selection, execution, hand-over, implementation, and manufacture, with all the in-between milestones, and checks and balances.

You can already see by this very simple list of major tasks that not only does it appear to be rather complex—and it is complex—but also time consuming. This is one of the reasons why companies started to open up and compartmentalize the individual elements of a project, thereby allowing verification of the needs-versus-capabilities status within the company. If the needs and requirements surpass the in-house technical capabilities, by doing so it becomes easier to identify potential partners, so-called know-how providers, wherever they can be found and brought to good and especially rapid use. My first basic, important answer to the question: "Why do we do open innovation?" is simply: "We want to accelerate the process of driving an important project to its successful end." There is obviously more to it, but this is a first, crucial finding: "It's the speed, stupid!"

It is like someone in your company has put out the mantra: "Speed isn't everything, it's the only thing." I am particularly emphasizing speed because it is not normally discussed as such a high priority.

I can add a personal mantra of mine, which goes like this: "If we had been faster at getting our projects as products or services into the marketplace, we wouldn't need patents, we would simply be ahead." I do understand that there are also other intrinsic values to patents, which are often used in the work of financial analysts when it comes to valuing a company, where number of patents per year, or number of patents per year and per 100 (or 1000, etc.) number of employees play an important role.

However, based on long-standing personal experience in the food industry, I can say with conviction that patents can sometimes even become counter-productive, because the technical community believes that by having patented something—a technology, a process, and ingredient, a usage, ideally combinations of these—they now have time and leisure to subconsciously slow down the process of development. I do realize that I have just made two enemies: the patent experts and lawyers, as well as many in the development community. Be that as it may, you should give these arguments some thought and a fair chance to be analyzed and discussed; if there was no patent protection (I am not speaking about other intellectual property protections of all kinds, such as marketing plans, trade secrets, recipes, etc.), how would you go about your development process? I can guarantee you that one of the likely answers will be: let's speed up the process and be in the marketplace as quick and as early as possible. There are certainly other answers as well, which might have to do with secrecy, non-disclosure, and maybe certain limitations on how much your company is willing to invest in your concept idea and run a fully fledged project in the absence of protection through a patent or a number of related patents. This is not a call against patents, it's simply a call to become faster in our development work, including all aspects of consumer insight, marketing plans, procurement strategies, manufacturing optimization, and sales campaigns.

Speed, in development as well as in execution, largely depends on the availability of two types of resources: the number of experts available in the company, and money. There is, however, one more over-arching pre-requisite to succeed, namely the willingness of your company to deploy these resources appropriately for the benefit of very few, very promising projects. Having fought a never-ending battle with managing large numbers of projects myself for my entire time in the food industry, I can say with some expertise that companies typically spread resources over a large—too large—number of projects, yet constantly seek to reduce the number of projects. If it wasn't a serious topic, this could almost be the stuff that stand-up comedians use in their acts.

There are several reasons for both directions, i.e. increasing as well as decreasing the number of projects. Increasing the number of projects in theory leads to a better distribution of risks, whereas reduction of the number of projects leads to greater efficiency. One of my former colleagues once formulated this in a rather cynical, yet realistic way: "The reason why there is such a large number of projects in our organization can simply be explained by the possibility for each one to better hide behind so many projects." Yes, as already said, it sounds cynical, but from many personal observations over many years I can say with some authority that this formulation is as close to the truth as one can possibly be. The counter-current to this is the reductionist approach: let's reduce the number of projects. But how do you do this, how do you even begin to approach it, and what type of mechanism or protocol leads you to a successful and efficient, result-driven reduction in the number of projects in any given situation? I do not have a simple, not even a complex, answer to this, as this is an ongoing movement, almost like a sine wave on a time axis.

I have seen reduction activities of many kinds, the most popular being to group similar projects together under one big project heading and calling all former projects "jobs" or something similar. This has, of course, not reduced the number of activities, neither has it freed up any expert resources or money to focus on the most promising and rewarding projects. There are many attempts undertaken in the food industry that I am aware of, and that drive towards better structuring of the development process, particularly the first concept selection process. In the previous chapter I discussed in some detail the tools at hand that help to fill the start of a development pipeline as, and this is almost a contradiction to what I have just described, there seems to be some void, especially in the early stages. This contradiction can be explained very simply: whilst organizations struggle to grasp the next great innovation and whole-heartedly support it, they rather put emphasis, i.e. people and money, behind ongoing, somewhat promising activities that have a very long life cycle and seem to go on "forever"; the accumulation of such development activities is the main culprit for the large number of projects that every food company apparently encounters, and is not too happy with. I personally must have come across the "world champion project in longevity": It started sometime in 1955 and was still active to some degree around 2005, literally celebrating its 50th anniversary. I do not want to be more specific here, out of respect for people who more recently have been involved in this project. I can, however, add that I was personally working on this project in the 1980s; mea culpa!

So, what's my point? Very simple: despite all the great project development tools, including checks and balances, milestones, tons of meetings, and even more meetings to discuss development advancements, it seems to be a lost battle when it comes to really controlling and properly managing the number of ongoing projects in the organization. Therefore, it is best to change direction and go outside. This sounds like an admission of defeat, but it's not! By doing this you are showing an important action in street-smartness: "Howl with the wolves," or, in other words, look for expert allies outside and create partnerships. Partnerships are based on open innovation principles, but they go further. It's not about simply putting your problem out into the world and waiting for answers, it's not even actively scouting for answers to your problems, but it is all about what the word actually means: search and become partners, behave like partners and treat each other accordingly. It's this important recognition that makes the important step from open innovation to innovation partnerships.

Let me conclude this section with a quote from an article I wrote together with my co-author Sam Saguy in 2009:

> The transition to OI (open innovation; the author) is not an easy task. Nestlé's history reveals three remarkable facts: (1) the company was established more than 140 years ago, (2) it has been exceptionally successful, and (3) nearly all of its flourishing innovations came from the inside. These facts, however, should also be the most important reasons for the need for change, especially when it comes to the source and origin of innovation. Nevertheless, it took some work by dedicated and committed people to change the company's approach to innovation. Nestlé's (then) new innovation strategy focuses on conducting business with its partners.

> This is termed Innovation Partnership, and it drives the journey toward what already is described in detail elsewhere as OI (Chesbrough, 2003) and Open Business (Chesbrough, 2006).

> The Innovation Partnership paradigm expands the OI definition, namely, a new way to collaborate in all areas of discovery and development with external partners who can bring competence, commitment, and speed to the relationship. The key to achieve this major goal is to accelerate necessary innovation and take the burden of resources and time pressure off the shoulders of a single partner.

(Traitler and Saguy, 2009)

THE CREATION OF THE INNOVATION PARTNERSHIP MODEL

Entering into a real partnership is easier said than done, otherwise there wouldn't be so many partner agencies out there. What is valid on a personal level is also of relevance when it comes to corporate partnerships: finding the *right* partner, finding a partner with similar values and expectations, and a few more criteria which I will discuss in some detail in this section. I have discussed the "why" of innovation partnerships in the section above and want to expand on the question of "how" in this section. There is no easy and simple answer to the question of how to do this, how to enter into an innovation partnership, and especially how to identify the right, the most appropriate, partners.

Let me recount the story of how we did this in Nestlé, beginning early 2006 with the first definitions of how innovation partnerships should be defined, structured, and executed within the large organization of a food company. Open innovation simply did not exist within the company, at least, if it existed in one form or another, it was not defined as such. The company probably did anything between 5% and 10% of its development activities with the outside world, most of it with academic institutions, and a sizable chunk, maybe half, through clinical studies in the area of nutrition, especially infant and health care nutrition. At that time, very few projects in real open innovation, let alone innovation partnership mode were carried out. Thus, for the company it was really new territory, or maybe not. Having personally been in charge of Nestlé's global packaging activities for more than three years, I came across this very interesting, company-atypical situation of doing large chunks of packaging development together with external partners, either really bespoke development or at least adaptation of stock packaging to the needs of the individual businesses. There were container projects, closure projects, material selection projects, easy-opening projects, packaging design projects, haptics improvements, and graphics and printing optimizations of many kinds and for every market and business. A very rough estimate led us to conclude that Nestlé probably did 2/3 to 3/4 of its packaging projects with outside partners.

So, here we were: approx. 70% of Nestlé's packaging activities were already done in some kind of partnership mode, whereas all other, product-, ingredient-, and process-related activities were mostly done inside, probably more than 90%. The conclusion from this was an easy one: if it works for packaging, it should also work for everything else. The reader may have discovered the catch already: people, or more specifically, numbers of resources available inside the company in the related areas. When analyzing the situation, we discovered that there were approximately 15 times more "all other activities"-related resources than packaging experts in the company. This was the biggest obstacle to overcome when we developed the answer to the question: how to establish open innovation and ultimately innovation partnerships as the way of doing development and business in the Nestlé company?

There are two aspects to this question: one is the purely internal, namely to convince the company people of the usefulness and efficiency of this new approach, and the other is linked to identifying and involving the external partners in the most successful way. Everyone who works in a large food company must have personal experience when it comes to the difficulty of promoting and selling new ideas to colleagues or bosses. Very often, you will find either outspoken opposition, or even worse, indifference. There are a few tips that can be helpful, which I will list in the following way:

- Prepare the plan and your arguments well, the best you can
- Remain simple and understandable in your arguments
- Build your group of allies beforehand
- Have the timing right
- Don't get discouraged by failure
- Listen out for the word "interesting"; everything is interesting, yet it's the "kiss of death."

There are certainly more tips out there, more personal ones that you have tried successfully, but my list above is already a rather complete one, which, if properly and "street-smartly" applied, should bring you some successes.

On the topic of selling new ideas to your company, this list should already be a good start. It has helped me tremendously, when, together with one or two colleagues, we prepared to sell open innovation and especially innovation partnerships to the Nestlé Company. We prepared our plan and our arguments during an almost gestational period of nine months and were able to convince layer upon layer of peers and bosses, up to general management, and got the whole thing approved. General management had to be involved, as the establishment of innovation partnerships was a strategy change for the company and therefore had to be approved at highest level. Do not be afraid to go to the highest authorities of your company in order to get approved for any idea with strategic connotations: it is ultimately very, very helpful. Here we were, a new strategy, a great plan, everything laid out and ready to go. The really critical step still lay ahead of us: to put it into living action! This is where the other aspect of "how" comes into play, namely to engage and convince the best possible external partners to collaborate with your company in ways they have never done before. There are many aspects that need to be considered here, and I shall discuss these in some detail in the following paragraphs.

It probably goes without saying that when you want someone to do something for you, you have to tell him or her first what it is what you want. I repeat this truism here, because I have personally lived many occasions, where colleagues in the company invited external experts to give talks, to show what they could do, to discuss their competences, without ever having told those external people what was potentially expected from them. It was like: "Tell us what you have and when I see something potentially interesting I shout Bingo." Maybe you have never experienced such situations, but my suspicion is that it happens all the time. And, that's not the way it should be. The very first thing one needs to do when wanting to work with external expertise providers, is to determine one's own needs and requirements, and that's exactly what we did at Nestlé at the onset of our innovation partnerships. I literally asked each individual business and business unit in the company to write down their needs and requirements for ongoing and future successes. The underlying hypothesis was that every consumer opportunity needs specific technical, scientific, and nutritional building blocks in order to succeed in the marketplace. Some of these building blocks were available inside the company, great! Others were not and those would be the ones for which expertise would either have to be developed inside—which would be costly and time consuming—or would have to be identified with external expert partners. It was understood that there could still be an important, mainly strategic reason, to build the missing competence inside, and that had to be fully respected. However, more often than not, it was decided that the needs and requirements list was better fulfilled by going outside. Receiving the requested information from the businesses and business units was not trivial at all. There was much suspicion throughout, but over time—it took more than a year—we got our act together and came up with a very complete document together. We organized this in two parts; the first part was business by business, while the second part re-grouped all listed needs and requirements into three sections: ingredients, processes, and packaging. Although it took more than a year to complete the document and it came in several iterations, we very early on and with what we had available then, turned to potential outside partners, which we not only grouped based on the stage of maturity of the expected innovation input from the outside, but also developed a short list of key criteria that would help particularly in the selection of partners amongst the company's key suppliers.

Figure 11.1 shows these criteria, as well as two, randomly selected weightings. I elected to show it this way to indicate the flexibility that we had to develop when we were dealing with outside partners, and when it came to organize such partnerships between a select group of partners, the company's businesses and business units, as well as the internal R&D community. It was not like squaring the circle, but it came pretty close. Figure 11.1 depicts five criteria, as well as possible weightings of such criteria:

- History: There is already successful ongoing business and collaboration
- Size: The partner has resources enough to develop and ultimately manufacture
- Need: The partner has a unique position
- R&D capabilities: The partner shows substantial creativity and innovation
- Cultural fit: We have similar values.

It is important to understand that there is no scientific solution to how the weightings are done, but they are useful in order to better understand what's really important and what may

Criteria for Innovation Partner Selection

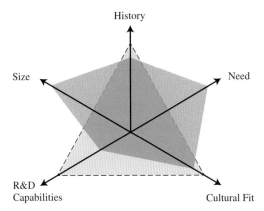

Figure 11.1 Partner selection criteria and two possible examples of weightings.

be less important when it comes to the most appropriate selection of potential innovation partners. Once the selection of the most useful and fitting partners is done, one is almost ready to enter into action, i.e. start the partnership. There are two more important areas to cover, before you can go ahead. First, because you are about to share important and sensitive information when communicating the list of needs and requirements with the selected partner, you have to make and sign the most watertight non-disclosure (NDA) or secrecy agreement, ideally a mutual one, and ideally for a duration of at least three years, better five years, but not longer. Extensive durations of NDAs do not really make sense, as they will eventually hinder your own company and can thus become counter-productive. The second element is gaining specific know-how on the selected partner's detailed competences, ideally organized in a similar fashion to the needs and requirements list, namely structured into ingredients, processes, and packaging. Once these two last points are successfully solved, you can proceed to the next step, which is sharing your list of needs and requirements with the selected innovation partner. There are two schools of thought, though.

The first one, of which I am a supporter, shares the entire needs and requirements document with the selected partner, thereby showing the partner everything and helping them to not only perform in areas in which your needs and the partner's competences overlap—the hot spots of collaboration—but moreover helps the partner to potentially develop into nearby areas of competence that might be of great value for future collaborations. The other school of thought prefers to limit the content of needs and requirements strictly to the overlapping hot spots and restrict any partnership to the here and now, what's available and what can help to solve a need as quickly as possible.

There is probably no easy answer as to which of these two viewpoints is better; it is more a question of what your company culture allows and also how far you are ready to open up and receive openness back. From personal experience I can say that we fared rather well by opening up entirely, because receiving openness back is not only reflected in the quality of the collaborative work, but also in the readiness of the partner to invest more in present and especially future collaboration, which ultimately results in more expert resources being put on collaborative projects. I can say with authority that openness directly pays off through the number of people resources that your innovation partner will provide. This is the real strength of the partnership approach, namely the access to expert resources that you as a company will not have to hire and train for three or five years before they can be used in complex projects. The partner already has them and throws them into the collaboration.

Another extremely important step in establishing innovation partnerships is to properly define the innovation landscape in which this approach is going to be embedded. We structured this landscape into three distinctively different parts, defined by the maturity of expected innovative solutions; we therefore divided this into:

- Early stage
- Mid stage
- Mature stage

We, somewhat randomly, decided that these stages were occupied by universities (early), start-up companies and inventors (mid), and suppliers (mature). Figure 11.2 shows the context as a simple overview.

The figure not only describes the three possible stages of maturity to be expected of innovative solutions coming from different partners, such as academia, start-ups, and inventors, as well as key suppliers, but it also shows how any collaborative activities are being financed by the company. As far as universities are concerned, all collaborative partnership activities are directly financed through the R&D budget of the company. Inventors and start-ups, if deemed to be relevant and promising are being financed from a different venture capital fund that most food companies set aside for such purposes. The third area of mature innovative solutions coming from key suppliers, who have been selected to become partners,

Typical Innovation Partnerships Stages

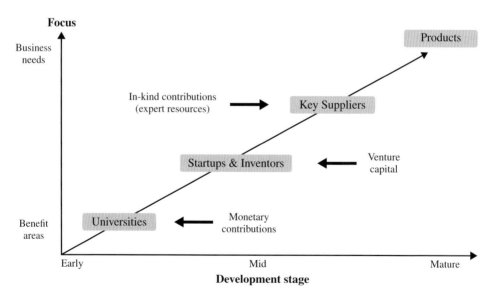

Figure 11.2 Stages of innovation partnerships (from Traitler and Saguy, 2009).

is the most directly and speedily rewarding one and is "financed" by in-kind contributions through expert resources from both sides, the company and the partner. One additional aspect of this figure is depicted on the y-axis: the more early stage the development is, the more generically it has to fit in a set of pre-defined benefit areas of the company. On the other hand, the closer we get to a possible execution of the project and subsequent launch, the more business input has to be given.

There is one more important, slightly philosophical thought that needs to be considered, namely the steps of engagement in a partnership, probably any partnership. There are really a number of indispensable dialogue steps that one goes through. The first, and probably most important, one is the establishment of trust: everything is built on trust and trust is earned mutually, hence the importance of some of the selection criteria, such as cultural fit and history. The next step consists of building goodwill in both, your company and the partner. This is not an easy task, as the "empire may strike back" at any moment. You will always find a varying number of people in your organization who would rather sit on the entire value chain (or value maze), and would reject at every possible occasion this approach of opening up, even if the partnership is extremely controlled and is probably the closest to open innovation. Beware these people and go forward by first building goodwill with those who are willing to support you. You can never convince the hard core of nay-sayers anyway, so don't waste time trying. Thirdly and ultimately the main, if not the only, reason for doing all this is to create additional and sustainable value for your company, as well as for the partner company.

Let me just very briefly give you a few facts and numbers to better illustrate, how the innovation partnership model that was developed by Nestlé worked out over time. Once we had the basic plan, the master plan for innovation partnerships worked out as:

1. We started by defining our own needs and requirements, identified by the businesses and business units.
2. We defined the innovation landscape: early, mid, mature, and defined who would be relevant partners at these stages.
3. We defined a number of key selection criteria for our key suppliers to potentially become innovation partners.
4. We asked our first selected partners to define and share their competences, structured into ingredients, processes, and packaging.
5. We signed mutual NDAs with the first 10 selected innovation partners in year 1 of the partnership activities; this number increased to 15 at the end of the second year, and approx. 18 after three years. There was a slight fluctuation in partner companies because we continuously checked their relevance versus business success.
6. We established a global network of Innovation Partnership Champions within the R&D system, by solely using existing resources.
7. Each partner company designated a key technical contact to become their go-to person. Unfortunately, this was not consistently done and there was a "mixed bag" of technical and business contacts.
8. We shared our own needs and requirements with these partners and kicked off the partnership activities, especially through *IdeaStores*, but also other direct development engagements.
9. In order to avoid that every set of ideas was jointly developed in such ideation sessions or other types of engagements, we, together with our partners developed a

so-called "Master Joint Development Agreement" (MJDA), which consisted of two main parts: terms of confidentiality, including each other's affiliates, and secondly, detailed project description, simply to be filled into the MJDA whenever necessary.

10. We also conducted so-called "top-level" meetings with each partner on a yearly basis. Typically two members of the executive board of our company, as well as several highest or high-ranking representatives of the partner company participated in these half-day meetings. Venues alternated between the two sites. These meetings were an important part of the checks and balances to monitor the efficiency of the ongoing partnership activities.

11. The 15 partner companies contributed more than 130 FTEs with a direct value of approx. $35 m, depending on how your company calculates the cost of one FTE.

12. At the end of the first year of our partnership activities we had already reached an average of 70 active projects with these partners. Most of the projects were on a fast track and we started to reap the benefits of launches based on these projects, as of the end of year 2.

In a recent meeting with Nestlé's Innovation Partnership Manager I learnt the following, in my eyes extremely smart new additions to the model:

- All selected partners are not just external expert resources of a certain value, but are treated as if they were an internal R&D resource, with all pipeline and project management considerations.
- The lists of partner competences are not only established by the partners and taken as such, but are continuously checked by the internal expert networks: trust is good, control is better.
- The new innovation partnership approach now insists that all partner contacts have to be technical people and not business people. This is not to say that there are no partner business contacts anymore, but these are to work directly with the business and procurement side of the company.

(S.Dobrev, Nestlé IPM, Personal Communication, October 2013)

Let me close this section with a few comments on the two other stages of maturity, namely early and mid stage. By applying the rigor of the master plan of innovation partnerships to not only key supplier companies, but also, albeit modified, to inventors, start-ups, and universities, we achieved a very structured, coherent and complementary approach to the entire external innovation landscape, thereby avoiding duplication of activities and ensuring the entire program was driven by the two main drivers: speed and results.

HOW TO DEAL WITH INTELLECTUAL PROPERTY IN INNOVATION PARTNERSHIPS

For those readers who, on an ongoing basis, deal with external collaborations, the words "intellectual property" (IP) have most likely almost become non-words. I remember times when I had to negotiate for 18 months with a university that did not want to budge over the question of who would become the owner of the IP of an as-yet non-existent collaboration. Part of the problem was of course based on the fact that my company wanted to own the

IP absolutely; the mantra of these days, not so long ago. On the other hand, the university did not want to budge either, so the conflict was pre-programmed. It almost ended in a stalemate, but ultimately we found some sort of compromise, by negotiating an up-front annual "success fee" that was to be added to the basic costs of the project. I do vaguely remember that not much earth shattering came out of the collaboration, however we lost time, trust, and nerves, none of which could really be recovered, unless it was the nerves. By the way, this was not the only frustrating case, there were many like this, and maybe even the majority of IP discussions ended very unsatisfactorily.

There was, however, a very important lesson from of all this: there must be another, simpler, more successful way of dealing with such IP questions. When developing the innovation partnership model, I also developed a new approach to sharing IP between two partners, in actual fact a very simple one, based on giving up the formerly iron-clad position of "we must own everything." Once this entered the mindset of my colleagues, things came together very quickly. What do I mean by this? It's very simple and it's got to do with the definition of owning. The mindset change was from, for instance, owning a patent to owning the application of such a patent. Let me give you an example: a flavor ingredient partner jointly develops a set of molecules by smart extraction, which is designed to substantially improve the taste profile of one of our products. Now, we progressed in very simple and pragmatic ways, by defining the ownership: the flavor partner owns the set of molecules and a patent that protects the-know how, but also defines the different possible fields of application. The food company, on behalf of which the development was performed in the first place, and which has also put in-kind resources into this development, subsequently receives the following freedoms:

- Exclusive, license-free application of the said set of molecules,
- In a negotiated number of product areas,
- For a negotiated geography, i.e. in which markets, and
- For a negotiated period of time, typically +/− 3 years.

The duration is normally volume dependent, i.e. duration can increase with increasing volumes, but should in all cases be finite, simply because through selling the set of molecules to third parties, the future price of the ingredients will become more affordable because of higher overall volumes.

There could be an important reason why some of the above may not work, especially the exclusivity, that doesn't necessarily have anything to do with the partnership approach, which may be caused by the respective market positions of the two partners: if they are number one in the category, anti-trust considerations might force the partners to open the development up and invite third parties to participate as well. This may sound strange, as it might cause the decision to not even enter into a joint partnership development; however, there could be strategic or business reasons why such a development could still be carried out.

I can say that in all the years of managing the innovation partnerships at Nestlé, I only had one such case in which it was decided not to enter into a joint development agreement. Overall, this new approach of splitting the cake into, as I would call it, "hardware" (for instance a patent) and "software" (the usage of the patent), made everyone's life much, much easier, sped up the process, and reduced unpleasant negotiations to almost zero. I can only strongly recommend adopting this approach in your company, irrespective of whether you are the primary company or the innovation partner.

TURNING PARTNERSHIPS INTO SUCCESSFUL AND SUSTAINABLE ENDEAVORS

I have talked quite extensively about the innovation partnership approach, about the rules of engagement, and best practices or simple rules as to how to start and successfully run them. Let me briefly expand on an even more important point: how to consistently succeed in this and help to sustainably improve your business. It has already been mentioned that there are always a lot of nay-sayers around you and that is especially true when you approach something new that takes many people out of their comfort zone. It becomes a further obstacle when your company is principally doing well and there is apparently no need for change. The nay-sayers and doubters will always tell you that it's nothing personal, but that your approach is not needed, is maybe not well thought through, doesn't bring any new and better results, is not the way that your company does business, and probably a few more such remarks. Don't become de-motivated, but remain steadfast. In order to do this you have to have two important elements on your side:

- Be extremely well prepared
- Be able to show undisputable successes

I do understand that the second is difficult to do when you start something completely new and disruptive to your company's typical way of doing business. On the other hand, one never, or almost never, starts at zero, so there should be some precedents to what you intend to do; dig them out and use them for your purpose, which is building on preliminary successes and working towards new ones. Nothing is a better basis for creating future successes than building on existing ones, past and present.

Let me discuss the first element, "being extremely well prepared," in a bit more detail. The error one could make is to confuse "prepared," just because it begins with the same first three letters, with "presentation." As a scientist, engineer, technologist, or representative of most other business disciplines, one learns, when starting a new project, any project, to first look what's there, what's existing in your company, as well as outside. Thanks to today's data storage capacities and programs, as well as networks, you should be able to really easily find what you are looking for: anything that can vaguely help you in your preparation. Don't do a study, do a digging and analysis job.

Once you have the elements together, you can start with the real preparation: take every bit of information that you have at hand apart and put it together in ways that will help your preparation: analyze and synthesize. Once you go through a first iteration of this process you will most likely find two things: first, what is still missing and second, a pattern of possible success stories. You then build on this and go into a second round of information gathering, analysis, and synthesis. Once you are happy with what you have achieved, you then can go and put it in whatever format is the most appropriate in your work environment. This could be a short report, a personal meeting with a flip chart, a short video in which you talk through the elements of what you found and intend to do, even by being asked in a kind of interview fashion, to be shared with your associates and peers. You can even do a slide presentation, if that's the way you do it in your company. In the end, it's always the personal engagement that counts and how convincing you are: the "shine in the eyes" of your argument.

All this is, however, just the very first step, really only the beginning of a long journey that is literally called "driving innovation through a complex organization to successful completion." jointly with expert partners and credibly so that the entire organization, including the innovation partner, continues to believe in it. This is by no means a trivial undertaking and requires substantial ongoing effort. Never letting the pressure go when it comes to successfully driving innovation partnerships through your company is the single most important success factor, cleverly combined with the above-mentioned elements of being extremely well prepared and being able to tell success stories.

By now you may have come to the point where you ask the very important question: "How many people does this require?" Point 6 of the action plan for innovation partnerships described above clearly describes the need for a strong network, a network that is ideally recruited from internal, existing resources. You can try and convince your management to hire additional resources, but I would predict that you will not succeed. Hence the need for a strong network that you strategically locate in places where your innovation partner can easily link in and, depending on your regional or global R&D organization, find your network members amongst existing R&D resources, but don't limit yourself to R&D. It is crucial that you make representatives from packaging, procurement, legal (IP), operations, and possibly marketing and finance members of your network with tightly knit and very specific job descriptions and expectations of what their exact role and anticipated results are.

A well-defined and continuously active and performing network becomes even more important when I consider the most recent twist to innovation partnerships described by S. Dobrev, namely the fact that the expert resources of the partner who work hand in hand with the company's resources are considered to be "one of us." By going down this route, great project management, well understood by both parties, is crucial, and respecting timelines, milestones, and expected results becomes critical for the sustained successes of innovation partnership activities. This is not a book about good project management, innovation pipeline, and administrative management tools, so I will not expand on these topics, as there is ample literature to be found. Let me just say this: if you want to be sure that you can successfully progress in partnership mode, all these elements need to be totally harmonized between your company and the innovation partner.

THE FUTURE OF OPEN INNOVATION AND INNOVATION PARTNERSHIPS

I have no crystal ball, nor do I read cards or anything similar. However, I have worked in the food industry long enough to know that it's very conservative. Most food companies that I know have not really embraced the term "open innovation," as "open" simply doesn't work for that industry. One could call it "open innovation under closed conditions" or something similar. It is not meant to be funny, it just reflects the reality and there are many reasons for this. One of the major reasons is that especially food companies do not deal with products that have high added value between raw materials and finished product. This means that every step in the value maze adds relatively little to the overall outcome, hence the desire to share that "little" with too many players is rather mitigated.

This ultimately results in the situation that the openness of the open innovation space becomes fairly controlled and semi-closed, as I would call it. This is nothing negative per se, as it means that you have to think a lot more about your problem upfront before you go anywhere outside or post it on your company website; the consequence is that the more closed option does not foresee such a site and most likely you will pre-select the potential solution provider amongst an already pre-selected number of the most promising. This was actually the starting point for innovation partnerships (INP), namely limiting the amount of internal resources necessary to identify the most promising solution provider by applying very stringent, formal, and detailed selection criteria laid out in the base plan, as well as in Figures 11.1 and 11.2. That was really the first evolution of open innovation in my company, going from OI, as described by H. Chesbrough and as applied by P&G in their C&D program, and moving on to the fairly controlled, yet highly effective mode of innovation partnerships.

Let me do some educated speculation and try to describe possible scenarios as to where OI and INP might go next. In Figure 11.3 I have attempted to depict such possible scenarios, without proposing that these are the ones that are going to happen. I started with closed innovation or "CI" and, i.e. the still often to be found situation that a company does literally everything themselves when it comes to progressing in innovation. This is not to be confused with "vertically integrated," it simply describes the situation in which most development people still find themselves today, despite hefty efforts by many to change this. This is also the reason why I can see one of the possible future directions of OI and INP is to go back to square 1 or CI, for the very reasons that I have mentioned above, namely low overall added value in the various steps of the value maze.

Let me discuss in a little bit more detail the possible ways forward, as I see them. What is already happening in OI mode is crowdsourcing, which will be further discussed in Chapter 12. However, crowdsourcing could be the precursor to crowd innovating. It is happening at a very low level, but this type of open innovation, which is a multi-expert partner approach, could take on more speed in the future, simply because it is the next logical step to crowdsourcing. As already mentioned above, the integration of partners into the innovation activities could go one step further: not only are the expert resources considered to be "one of us," but they could come and join "us" in our own premises and be truly embedded, physically as well as in day-to-day ongoing innovation activities. By the way, the embedding could also go the other way: your expert resources relocate to the partner's premises

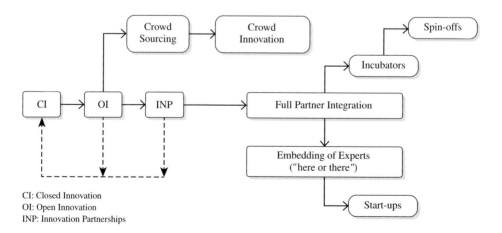

Figure 11.3 Possible scenarios for the future of open innovation and innovation partnerships.

and become a fully integrated team. One interesting scenario that I could see here, and that carries a certain danger for both parties, is the possibility that such activities will generate opportunities to create start-ups formed by the experts. A good thing would be to release such a very dedicated start-up into temporary freedom and, through right of first refusal contracts, get first-hand access to a potentially great outcome without having to finance the project. Yes, the resources are temporarily gone, but they would be anyway, even if they continued to work inside the company. This certainly merits some thought and probably lots of discussion as to how this could be optimized in order to achieve almost guaranteed results.

The other variant of full partner engagement could be the creation of incubators in the first place. It's not the same situation as just described above, because such an incubator would still fly under the wings of the mother companies and be a full member of their innovation-expert network. Incubators are no real love affair in food companies, especially not in the one I have personally worked for a long period of time. Although it has never really been tried, there are immediately many voices, especially from the top that, probably out of fear and not knowing any better, kill any attempt towards incubators even before it had a chance to be discussed. My feeling is that there may be a third reason: fear of loss of control. Especially top-level managers experience this quite a lot, as I have observed many times. However, Incubators would probably be a great way forward towards speeding up innovation in any size company, small or large, again especially in food companies. One possible outcome of the pathway through incubation is the likely creation of spin-offs, similar, but not identical to start-ups, as they typically have a head start and further financing should be easier, although still not guaranteed.

Let me quote from a yet unpublished study that my co-author I. S. Saguy and I conducted in 2009.

> *Business incubation is a process enacted by corporations, angel and venture capital organizations in order to facilitate and promote entrepreneurial and innovation processes. Incubators typically provide a variety of services, such as developing new technologies, marketing plans, minimizing time to market, building management teams, obtaining capital, etc. After the incubation period, it is intended that incubation ventures graduate to either become independent and self-sustaining businesses or be assimilated into the corporation. Generally, incubators have been recognized to provide an excellent opportunity for promoting innovation and making significant contributions.*

One of the major conclusions was:

> *Most large corporations are struggling especially in managing the incubators and bringing in its people while simultaneously searching for ways on how to provide the challenge and free working environment not to stifle their innovation.*

> (I. S. Saguy, and H. Traitler, Internal Study, 2009; unpublished)

CONCLUSIONS

In this chapter we discussed and critically analyzed questions around open innovation (OI) and innovation partnerships (INP). We talked about transition from OI to INP and how INP was created and applied, including the master plan for INP. We discussed partner selection criteria, the stages of maturity of partnership collaborations, and, very importantly, how to best deal with IP issues. Moreover, we discussed the questions how to make OI and INP sustainable and durable in your organization, and finally tried to elaborate on possible future scenarios in the field of OI and INP. We can conclude the following points:

- Since P&G and H. Chesbrough created the models behind Connect & Develop (C&D) and Open Innovation, many companies have attempted, more or less successfully, to use these models to their benefit. Ultimately, all development work needs two major elements: resources and time. Innovation partnerships were born out of the need to manage both of these and still deliver speedy and innovative results.
- The creation of the INP model was largely due to the facts that there was this need for speed by tapping into external expert resources, without going the whole way and hiring a large number of internal experts who could manage the approach and check the validity of external proposals.
- We discussed and analyzed the basic requirements for entering into successful partnerships, one of which is a solid and functional list of selection criteria, serving as the basis for selecting the most promising innovation partners from a list of, more often than not, already existing suppliers to the company.
- We also presented and discussed the innovation landscape into which the INP activities were to be embedded and how these are supported resource-wise, by direct financing, venture capital, or through in-kind existing expert resources.
- As the centerpiece of the creation of the INP approach, we discussed and analyzed in detail the master plan that has to be put together in order to create successful and sustainable INP activities, followed by a list of 12 key steps to be taken in order to get started.
- The INP approach has evolved since and now added three additional, very smart criteria:
  1. External expert resources are treated like internal ones and follow the same project management and other procedures
  2. Partners' competences are checked by internal experts
  3. Partners' counterparts have to have a technical background.
- We discussed questions linked to potential issues of intellectual property ownership and recommended to move away from: "We want to own everything" and migrate towards: "We want to own the application, for the agreed upon period of time, for the agreed upon geography, and for the agreed upon groups of products."
- We discussed and analyzed the need to turn INP into successful and sustainable endeavors for you as well as your partners, and emphasized such elements as being extremely well prepared for critical discussions, being able to put together success stories, and building a very diverse, supportive, and active network.
- Lastly, we discussed the question regarding the future of OI and INP and presented different possible scenarios, amongst which we included the one going backwards towards closed innovation, but more promising were the scenarios forward towards "crowd innovation," full partner integration followed by either the embedding and possible start-ups route, or, alternatively, daring the creation of incubators, potentially leading to spin-offs.

TOPICS FOR FURTHER DISCUSSION

- Irrespective of the nature of your work (technical, procurement, marketing, etc.), and on which side you work (competence requester or competence provider), what is your view on how OI or INP have shaped the innovation landscape in past years? Critically discuss pros and cons, and synthesize your conclusions and how this best fits your organization.
- What's your company's approach to open innovation? Please get some insight and compare it to other companies' approaches, especially to the above-described partnership approach. Critically compare, analyze, and discuss, especially the usefulness of open innovation in general and how this might have affected business in your company.
- As a representative of a partner company, the "knowledge provider," what is your experience of the approach of collaborative partnership projects? Collect good and bad examples, and discuss these in detail within your teams. Discover and propose best ways forward to maximize efficiency of the approach.
- Discuss the element of fear of job loss that you might encounter when operating an OI or INP approach: "If my company finds external experts, am I still needed?" List and discuss the most important reasons why you and your work are even more needed in OI and INP mode.
  Note: I have, on purpose, not discussed this topic in the chapter above, because I wanted the readers to find out themselves. Rest assured, there are many good reasons why you shouldn't be afraid!
- Discuss and analyze speed of progression and completion of projects, and, based on your own experience, what elements, other than OI or INP you can find to substantially shorten project development times. Compare and critically analyze each of these elements and build a comprehensive approach to "speeding up development."
- Discuss and analyze the proposed way forward in dealing with IP issues when operating in OI and/or INP mode. Could this work for your organization, and if not, what is your solution to this important topic?
- To close the loop to the first discussion point above, how do you see the future of OI and INP? Critically discuss this topic, especially the options of going back to CI, and whether you see this as a threat or, conversely, a desirable direction towards "back to basics."

REFERENCES

A.G. Lafley , R. Charan (2008). *The Game Changer*, Crown Business, Crown Publishing Group, New York.
H. Traitler, S. Saguy (2009). Creating successful innovation partnerships, *Food Technology*, 03.09, 24–25. www.ift.org.

12 What can the food industry learn from Silicon Valley?

Without change there is no innovation, creativity, or incentive for improvement. Those who initiate change will have a better opportunity to manage the change that is inevitable.

William Pollard

INTRODUCTION

From innovation partnerships to networking is only a small step, if a step at all. The role of networking will be emphasized in this chapter and networking experts will be reporting and sharing their expertise, and will discuss the results they were able to achieve. Networking also encompasses the need for interaction that goes beyond so-called "events" and the like, but calls for serious joint engagement in person-to-person meetings, brainstorming sessions (yes, good old-fashioned VanGundy, and Osborne and Parnes) and panel or round-table meetings with public interaction are important elements of the new, comprehensive, and successful networking. Also, the rise of entrepreneurs—the key ingredient of Silicon Valley—but this time the rise of entrepreneurs in Silicon Valley within the food industry will be discussed in this chapter as well, with examples and relevant results, both in qualitative (level of engagement) and quantitative (added value to the business) ways.

HI, I AM A CONNECTIONS EXPLORER

Hi, my name is Birgit Coleman. I am a Connections Explorer. Believe it or not, my job title is "connections explorer." What makes somebody a connector? Malcolm Gladwell defines it in his book, *The Tipping Point: How Little Things Can Make a Big Difference*, as the following: *Connectors*, are the people in a community who know large numbers of people and who are in the habit of making introductions. A connector is essentially the social equivalent of a computer network hub. They usually know people across an array of social, cultural, professional, and economic circles, and make a habit of introducing people who work or live in different circles. Malcolm Gladwell characterizes these individuals as having social networks of over 100 people. To illustrate, he cites examples such as Milgram's

Food Industry Design, Technology and Innovation, First Edition.
Helmut Traitler, Birgit Coleman and Karen Hofmann.
© 2015 John Wiley & Sons, Inc. Published 2015 by John Wiley & Sons, Inc.

experiments in the small-world problem, the "Six Degrees of Kevin Bacon" trivia game, Chicagoan Lois Weisberg. Gladwell attributes the social success of connectors to the fact that "their ability to span many different worlds is a function of something intrinsic to their personality, some combination of curiosity, self-confidence, sociability, and energy."

Six degrees of separation

Some of Gladwell's analysis as to why the phenomenon of the "tipping point" occurs is based on the 1967 "Six Degrees of Separation" study by psychologist Stanley Milgram. The study found that it took an average of six links to deliver a message. Of particular interest to Gladwell was the finding that just three friends of the stockbroker provided the final link for half of the letters that arrived successfully. This gave rise to Gladwell's theory that certain types of people are key to the dissemination of information.

The strength of weak ties

Mark Granovetter, a sociologist at Stanford University who has created the theory of "The Strength of Weak Ties" argues that all "bridges" are weak ties and that weak ties are more likely to link members of different small groups than are strong ones. So, it is more likely that your random contacts can connect you or bring you closer to your destiny than your best friend or circle of friends. The fewer indirect contacts someone has, the more encapsulated he or she will be in terms of knowledge. Those to whom we are weakly tied are more likely to move in circles different from our own and will thus have access to contacts and/or information different from that which we receive. It is actually remarkable that people receive crucial information from individuals whose very existence they have (almost) forgotten.

The irony in my destiny is that I arrived at the end of February 2001 from Vienna, Austria knowing nobody, and today I am a Connector and Connections Explorer for Switzerland. Since December 2002, I have worked for Swissnex San Francisco—at that time still called the Swiss Science & Technology Office. The mission of Swissnex San Francisco is to connect the dots between Switzerland and North America in science, education, art, and innovation. It is a public–private venture and an initiative of Switzerland's State Secretariat for Education and Research, managed in co-operation with the Swiss Federal Department of Foreign Affairs as an annex of the Consulate General of Switzerland in San Francisco. Swissnex San Francisco is part of a network of knowledge outposts in Bangalore, Boston, Shanghai, Singapore and most recently Rio de Janeiro, along with individual science counselors around the world, working to connect the dots. Through 80 to100 public events per year and numerous study tours across industries, it highlights the best of Swiss and North American ingenuity and discovers and creates opportunities for networking among diverse group of professional contacts in the San Francisco Bay Area, and beyond. The team believes that unexpected encounters can lead to the brightest achievements, and it encourages the cross-pollination of ideas through the many connections made at Swissnex.

This is where I come in as a Connections Explorer and Connector to our Swiss stakeholders. I act reactively to their requests and I proactively suggest names I think they should talk to in order to discover opportunities for partnerships and facilitate these introductions. In addition to brokering the contacts, I also act as a "maven." *Mavens* are "information specialists," or "people we rely upon to connect us with new information." They accumulate knowledge, especially about the marketplace, and know how to share it with others.

As Malcolm Gladwell states, "Mavens are really information brokers, sharing and trading what they know." So, it is not just the connections I/we at Swissnex San Francisco broker, but also the content of or the information about the insights of "What's happening in Silicon Valley" that we communicate to our Swiss stakeholders.

Open innovation has gained increasing popularity in recent years as organizations recognize the value of connecting and collaborating with outside people. Sun's founder Bill Joy noted many years ago, that "most of the smartest people work for someone else." Similarly, as countries focus more and more on research, innovation, technology, and entrepreneurship, they are increasingly reaching out to and connecting with key innovation centers around the world and have set up "bridge organizations" to help build bridges between their home country and these innovation centers or hubs, like Silicon Valley.

To give you examples, in early 2007 I started working within Swissnex San Francisco with the Innovation Partnerships Group of Nestlé, the Swiss multinational nutrition, health, and wellness consumer goods company headquartered in Vevey, Switzerland. It is the largest food company in the world, measured by revenues. In the Nestlé case, Swissnex San Francisco got hired to explore potential innovation partners for Nestlé within the start-up and investors community, academia, and big corporations to enhance or transform a product company into a product and service company, and to connect with the consumer on an emotional, personalized, and transparent level. Even a powerhouse company like Nestlé needs trusted, well-established, and nurtured introductions, especially to the technology giants and universities that are housed in the San Francisco Bay Area and the US West Coast such as Google, Facebook, Amazon, Microsoft, Stanford University, University of California, Berkeley, Art Center of Pasadena, Caltech, etc., in order to fulfill this mission.

As a nice by-product, Swissnex San Francisco accomplished getting Nestlé into *Bloomberg Business Week*'s world's top 50 most innovative companies for the first time, in 2009 and 2010. How did we achieve this? It happened in untraditional ways by hosting at Swissnex San Francisco many public and private innovation round-table discussions with chief innovation representatives from various local corporations, such as McKesson Corporation, Clorox, Safeway, Chevron, etc. mixed with design and sustainability experts, entrepreneurs, and always with a special guest from the media—from Fast Company, to *The Wall Street Journal* to *Bloomberg Business Week*. It was that unconventional pull versus push approach with the media that gained Nestlé that necessary additional and innovation-targeted attention and traction that ultimately spilled over into the top 50 most innovative company ranking. A push–pull system in business, according to Wikipedia, describes the movement of a product or information between two subjects. On markets the consumers usually "pull" the goods or information they demand for their needs, while the suppliers "push" them toward the consumers. Another meaning of the push strategy in marketing can be found in the communication between seller and buyer, in our case between Nestlé (and the other roundtable participants) and the media, where latter had demands on Nestlé's information on best innovation practices. By basically opening our books and sharing with the media our insights of internal and external innovation challenges and best practices, we gained their continued trust by showcasing non-stop the company's commitment to responding to innovation demands in new, efficient, and personalized ways.

Another soft marketing approach was to share information about the brain–gut axes via a TED talk "The brain in your gut" by placing a Nestlé veteran and food scientist, Heribert Watzke, at the TED Global 2010 conference, thanks to Swissnex San Francisco's connections to Bruno Guissani, European Director of TED. In Heribert's talk we learned that we have about a hundred million neurons in our intestines, our "hidden brain" in our gut, and about the surprising things it makes us feel. TED (Technology, Entertainment, Design) is a global set of conferences owned by the private non-profit Sapling Foundation, formed to disseminate "ideas worth spreading." Since June 2006, the talks have been offered for free viewing online. On November 13, 2012, TED Talks had been watched one billion times worldwide, reflecting a still-growing global audience. Personally, the videos nourish my discovery-seeking brain with a little hit of dopamine in the middle of a workday. These are examples of why we Connectors and Connections Explorers qualify for our own profession and are in demand.

FORMAL APPROACHES TO INNOVATION PARTNERSHIPS

It is less and less a secret that big powerhouses need help with these innovation tasks, such as finding the right innovation partner for the company's new future that gets rapidly disrupted. There are a few ways in which corporations can tackle these challenges.

Singularity University

In 2013, Singularity University, the X Prize Foundation and Deloitte Consulting formalized such a process by launching a four-day executive education program: The Innovation Partnership Program (IPP). The immersive three-year partnership program prepares senior executives, from global corporations and Fortune 500 companies like Google, Shell, Qualcomm, and Hershey, to take their companies from "linear" to "exponential," by creating a culture of explosive innovation. The Executive Program enables corporate innovators to keep pace with accelerating change and understand how emerging technologies will impact their industry. Participants learn to recognize the growth opportunities and disruptive influences of exponentially growing technologies and how key breakthroughs in areas will impact their careers, companies, and industries in the coming years. This program is offered as a seven-day workshop or custom program for corporate groups.

Corporate venture groups and innovation labs

Corporations such as AT&T, Swisscom, SK Telecom (a South Korean wireless telecommunications operator), Citigroup, Citrix, American Express, major car companies such as BMW, Honda, Nissan, Daimler Chrysler, etc. also have their innovation labs/venture groups in the Palo Alto area in order to accelerate technology.

According to TechCrunch, Silicon Valley leads the nation with more than 50% of telecom venture investments in 2012, and its share is growing. The reason for this is that companies such as Google, Apple, VMware, and Salesforce have redefined the meaning of the telecom carriers business to include Internet, mobile and cloud, with wireless, terrestrial, and satellite transmission of voice and data. In turn, Silicon Valley telecom entrepreneurs have pivoted from capital-intensive hardware infrastructure with expensive and long telecom carrier sales cycles to less capital-intensive, software-based products and services with shorter telecom carrier sales cycles. Beyond attracting venture capital, the concentration of innovation has also attracted carriers to Silicon Valley. AT&T, Sprint, China Mobile, France Telecom/Orange, SK Telecom, and Swisscom are among representatives of the many large and small American and international carriers that have offices in Silicon Valley. All share the common goal of accelerating R&D and product delivery cycles with innovation acquired from start-ups.

For *AT&T*, with its AT&T Foundry innovation center in Palo Alto, the goal is to collaborate with tech leaders and start-ups to fast-track new apps and platforms.

American Express has created a $100 million investment fund to ensure it won't miss out on the next digital commerce opportunity, including start-ups that are focused on loyalty and rewards programs, personalized offers, location-based services, security issues, analytics, and online and mobile payments.

Citi CIO Deborah Hopkins and her Palo Alto team met with hundreds of start-ups. Their subsequent investments in Square, Jumio, and others helped Citi gain traction in the mobile-payment space.

The *Honda Silicon Valley Lab* (HSVL) was established in May 2011 in Mountain View as an open innovation lab for global Honda R&D, focused on information technology. The recent integration of Apple Siri Eyes Free into certain Honda and Acura vehicles is a recent example of Honda's effort to lead the industry in bringing new information technology into its products. In early 2013, HSVL established, with Evernote, a New 'Honda Innovation Award' for app developers. Evernote is helping the world remember everything by building innovative products and services that allow individuals to capture, find, and interact with their memories. This partnership will nurture and award the development of Evernote-integrated apps that make personal mobility more enjoyable and memorable. It is a six-month program featuring a series of in-person and virtual events that encourage developers from all over the world to create new apps on the Evernote platform.

In March 2013, *British Airways* (BA) also followed suit by launching the first of its "UnGrounded" innovation lab flights that assembles 100 Silicon Valley luminaries from Google, Andreessen Horowitz, RocketSpace, and elsewhere on board a flight from San Francisco to London on June 12, 2013 to devise a platform for connecting tech talent with big problems around the world. The plan is for the UnGrounded "innovation lab in the sky" flights to become a somewhat regular occurrence with different destinations. On the first one, though, the passengers addressed the problem of mismatches between where technology innovators live and where the problems they could solve reside. BA's help came from Eric Schmidt's Innovation Endeavors fund and the RocketSpace start-up accelerator, and IDEO designed the in-flight experience. Why did BA do this? TechCrunch disclosed the EVP of BA, Simon Talling-Smith's explanation that the spirit of innovation has changed. A lot of the activity is happening in the technology sphere and they asked themselves: "What could we do to play our part?"

Another out-of-the box innovation lab is *Hattery* in San Francisco, as shown in Figure 12.1. It is a venture fund and innovation lab with a culinary team dedicated to nourishing creative minds with healthy food and to bring all entrepreneurs-in-residence and staff, including partners around the table once a day. The *Hattery* fund invests at the early stage in user-focused, applied technology companies that promise to make a positive impact. It supports and grows their portfolio companies through: (1) offering space in a bright, airy San Francisco office to work, eating at the *Hattery Kitchen*, and share ideas with the team, advisors, and entrepreneurs-in-residence, (2) providing access to Hattery's team of designers, engineers, strategists, and financial experts and (3) seed funding.

Figure 12.1 The Hattery, San Francisco. (Copyright: Birgit Coleman, March 2013).

Bridge organizations

Like Swissnex San Francisco, other countries also have their innovation centers in the San Francisco Bay Area and their eyes and ears on the ground for their stakeholders. Swissnex San Francisco was among the first, by moving out of the Consulate General of Switzerland in spring 2003 and officially opening its doors in November 2004. Other strong and active country representations are the Innovation Center Denmark, Finnode (a global network of Finnish innovation organizations), Norway's Innovation House Silicon Valley, UpWest Labs (a Silicon Valley accelerator supporting Israel's Best Entrepreneurs), etc. Even though each innovation center has their own way of connecting the dots between Silicon Valley and their home country, our common goal is to accelerate innovation and have our home country be part of it.

BUT HOW DOES SOMEBODY BECOME A NETWORKER OR CONNECTOR OR A CONNECTIONS EXPLORER?

As the title "networker" gives it already away, through attending lots of networking events and being curious about meeting people and learning from and listening to them, you can populate and boost your own network. Then the "game" has just begun: your next step is to nurture your newly acquired contacts by turning them into trusted relationships by

Figure 12.2 Building your own net and connecting the dots. (Reproduced with permission of Susanne Kühlewein).

giving and sharing (and taking) within your eco-system/network and also to "track" them in their careers as these skilled employees move rapidly between competing firms in the San Francisco Bay Area and beyond (Figure 12.2). Software tools like LinkedIn, Xing, Facebook, BranchOut, etc. make the tracking task easier as people generally update their user profiles with every career move.

As you still may remember, I arrived in the US in 2001 and started working for Swissnex San Francisco in 2002, with a scant local network under my belt at that time. I stumbled into networking as a pure coincidence, out of necessity and survival in 2001 in order to connect with people for the sake of making friendships, but also to integrate into the foreign business ecosystem to hit the ground running professionally. People at any stage of their careers and in any type of job need a good dose of networking and the San Francisco Bay Area does not lack in offering these networking grounds. It is quite the opposite—you could network around the clock. Interestingly, there are a lot of soft facts—the numerous networking events with an apparatus of infrastructure—that make the San Francisco Bay Area what it is.

THE POWER OF NETWORKING: NETWORKING PRINCIPLES

In a white paper "Network! Network! Network!—How Global Technology Startups Access Modern Business Ecosystems," Antti-Jussi Tahvanainen, Research Institute of the Finnish Economy/Stanford University, Center for Design Research, and Martin Steinert, at that time with Stanford University, Center for Design Research, explain the three networking principles that are working behind the scene to unleash the power of networking:

Serendipity. The real power in a serendipitous approach is in enabling access to a much larger pool of opportunities than a purely planning-based approach could. Serendipitous exploration enables access to the unknown unknowns that simply cannot be targeted through planning because, by definition, people are not aware of the existence of these opportunities in the first place. As haphazard as this approach may seem, serendipitous networking is very purposeful and strategic. It involves a lot of explicitly dedicated time and strenuous work. Being strategically serendipitous implies that networking just for the sake of networking must be perceived as a fundamental and explicit part of work. It is an important complement that increases the probability of encountering relevant opportunities.

Flexibility. Serendipitous strategy emphasizes constant iteration of ideas, business models, and product concepts. It requires a great deal of flexibility and the ability to open up to new ideas at the cost of original ones. Feedback, critique, and ideas obtained from new contacts are seen as an opportunity for improvement, not as a rejection of concept.

Reciprocity. Serendipity lives off a culture of reciprocity. It is that shared ecosystem-wide understanding that, for an individual to extract benefit from a random encounter, everyone in the community needs to be willing and ready to provide it. Because of the prevailing culture of reciprocity in Silicon Valley, the start-ups found it relatively easy to approach new contacts and retrieve feedback and advice, as long as they were able to convey a promise of mutual interest. The interest does not necessarily have to result in immediate benefit. Providing someone a favor is considered an investment that might create returns at some unforeseeable time in future. It creates social capital that can be called upon when needed, possibly in a very different context.

SILICON VALLEY AND ITS ECO-SYSTEM

First, let me make some necessary introductions about Silicon Valley. Silicon Valley is the southern region of the San Francisco Bay Area in Northern California, USA, as depicted in Figure 12.3. The region is home to many of the world's largest technology corporations and thousands of small start-ups. The term originally referred to the region's large number of silicon chip innovators and manufacturers, but eventually came to refer to all the high-tech businesses in the area. It is today generally used as a metonym for the American high-tech sector.

Despite the development of other high-tech economic centers throughout the USA and the world, Silicon Valley continues to be the leading hub for high-tech innovation and development, accounting for one-third of all of the venture capital (VC) investments in the USA.

What makes Silicon Valley or the San Francisco Bay Area in general one of the leading hubs for high-tech innovation and development? It is the cross-pollinating eco-system

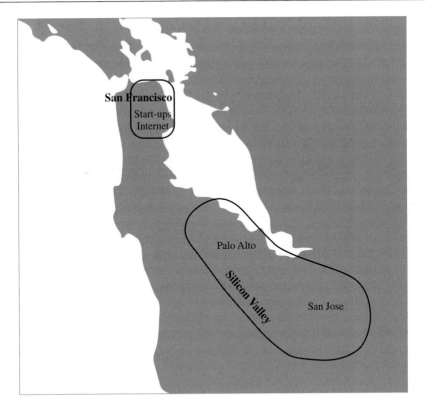

Figure 12.3 Map of the so-called "San Francisco Bay Area" and Silicon Valley.

that comprises components such as the investment community (VCs and business angels, friends and family) and start-ups (with its serial entrepreneurs) and large local corporations (such as Google, Facebook, Twitter, eBay/PayPal, Cisco, Intel, etc.), local top universities such as Stanford University, University of California, Berkeley, research institutes such as Xerox Parc, SRI, etc., numerous of incubators, accelerators, and co-working spaces for start-ups coming from abroad or locally seeking community and support. The lubricant between these components is a magic mix between many days of sun, open-minded people, and lots of networking.

In my 2012 Swissnex San Francisco's Nextrends Blogpost "Surveying the Incubator and Accelerator Landscape," I showcased the powerful eco-system Silicon Valley has for its entrepreneurs that feeds into the innovation pipeline of the big corporations. Start-ups are moving from coffee houses and garages to co-working facilities, incubators, and accelerators. The three categories vary widely and there is often overlap, but in general, they each provide a unique set of offerings to entrepreneurs and start-ups. By the end of the day, the start-up is buying itself into the network of the respective incubator and/or accelerator, as this will open opportunities and doors to potential investors, attorneys, business partners, and/or clients (Depicted in Figure 12.4).

The San Francisco Bay Area offers many *co-working facilities*, such as WeWork, The Hatchery SOMA, StartupHQ, 5M Project, and Mission*Social, among others. These attract

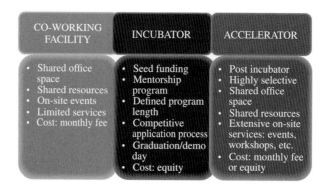

Figure 12.4 Table highlighting the differences between a co-working facility, an incubator, and an accelerator.

professionals who want to get out of their houses and engage with others in an office environment. Events and classes are held on the premises, but services are usually limited. Co-working spaces typically charge a monthly fee.

In comparison, *startup incubators* such as YCombinator, 500Startups, and TechStars are a facility or a mentorship program for early stage start-ups. Incubators have a competitive application process, and a company is accepted into the program based on its ideas and the team's potential. Each company stays in the incubator for a specific amount of time and then "graduates" at the end. Resident companies receive assistance refining their business ideas and preparing the product for market launch. In exchange, the incubators receive equity in the companies they help.

A *startup accelerator*, such as RocketSpace, Kicklabs, BootstrapLabs, Blackbox Mansion, i/o Ventures, etc. picks up where incubators leave off. If an incubator is similar to an undergraduate education, an accelerator is like graduate school. Accelerators focus on post-incubator, seed-funded start-ups on the cusp of rapid growth. Like incubators, accelerators are highly selective. They provide participating companies with a workspace, extensive start-up services and networking opportunities, with angel investors, VCs, lawyers, and leaders of Fortune 1000 companies. Accelerators typically charge start-ups a monthly fee, but they may have the option to invest in the companies they advise. Since 2005, Y Combinator, for example, funded more than 460 start-ups, including Reddit, Dropbox, Heroku, and Airbnb. Since 2010, 500 Startups seed-funded more than 260 internet start-ups, including Wildfire Interactive acquired by Google, as well as TaskRabbit, Twilio, and Mint.com, which was acquired by Intuit.

WHAT ABOUT FOOD AND TECH INCUBATORS/ ACCELERATORS/CO-WORKING SPACES IN THE USA AND THE SAN FRANCISCO BAY AREA?

A new group of food-based start-ups are applying tricks learned from the technology industry to grow a new wave of businesses to cash in on the growing "foodie" movement across the USA. Small food businesses—mom and pop operations selling goods at farmers markets, food trucks, etc.—have proliferated in the last few years, according to data provided by the National Association for the Specialty Food Trade. Both incubators and accelerators offer small food businesses the opportunity to grow within a nurturing environment, while defraying large start-up capital costs. While traditional accelerators have been aimed at Internet-based start-ups, food accelerators are a much newer proposition. Food businesses traditionally require thousands of dollars in launch capital and are subject to much more intensive government regulation. 500 Startups is nurturing mail-order food businesses, among others. The young Local Food Lab, soon-to-be Forage Kitchen, the 14-year-old La Cocina, and The Food Craft Institute (FCI) are great local examples.

Forage Kitchen will be an incubator and co-working space for artisan food businesses, but also a space for non-professionals to learn and cook with professional equipment. For $99/month, you get access to the kitchen, training, culinary support, and equipment. Interestingly, Forage Kitchen started as a Kickstarter project. Kickstarter is a funding platform for creative projects from the worlds of music, film, art, technology, design, games, fashion, food, publishing, and other creative fields.

The Local Food Lab in Palo Alto and New York City offers a six-week program focused on business training for the next wave of food and agriculture ventures. In order to scale their reach and business as well, the Local Food Lab is ramping up to become a pre-Kickstarter online platform where entrepreneurs can post or pin (like "Pinterest," with the difference of pinning your own creation) their own work portfolio and raise awareness and money that way (Krysia Zajonc, personal communication, October 2013).

La Cocina is a well-established kitchen incubator that helps low-income entrepreneurs, especially women, from the immigrant community to become self-sufficient doing what they love to do.

The Foodcraft Institute works to create and improve the viability of small- and medium-scale value-added food businesses in rural and urban America. It is a new educational institute that combines classroom and hands-on education to teach traditional food-making techniques alongside the entrepreneurship skills needed to turn those strengths into viable businesses. One course example covers how to own and run a farm-sourced butcher shop, where you can learn the keys to a great relationship with ranches, slaughterhouses, distributors, and customers. A close partner is the local *Belcampo Meat Co.*, which is a farm, a processing plant, a butcher shop, and a restaurant, and manages the entire production chain.

FOOD INCUBATORS AND ACCELERATORS OUTSIDE OF SILICON VALLEY

Most of other large urban markets such as New York and Los Angeles offer medium-to-large sized commercial kitchens. The concept of a kitchen incubator, also known as a culinary incubator, relies on the fact that FDA (Food and Drug Administration) and state regulation prohibit the sale of food that is not produced in a licensed facility. Culinary start-ups are unlikely to receive venture capital or bank financing, as profit margins are too slim and volatile for such a highly competitive market. Food products must be tested and tweaked over time before they are economically viable. Even once proven viable, the entrepreneur must navigate a complex network of regulation, packaging, and distribution before running a profitable enterprise. This entrepreneur often lacks a business background and an understanding of what is involved in the start-up process. A study of individual demand for kitchen rentals reveals that start-up costs and licensing complications are the two main deterrents to opening a private kitchen. Availability and reliability are listed as the two major deterrents for aspiring entrepreneurs. But despite the growing popularity and apparent need, food incubators and accelerators, they aren't everywhere: apart from outliers such as The ARK, most are situated in large cities or in dense suburbs. Replicating the food incubator model in rural areas and college towns is the next great challenge for the growing field, which the San Francisco Bay Area has already tackled with the above-mentioned examples.

The accelerator in Arkansas, *The ARK*, pays special attention to food. Due to Wal-Mart's and Tyson Food's headquarters being located inside the state, The ARK is specifically recruiting food startups who could benefit from close proximity to the agribusiness giants. The ARK offers recipients approximately $18,000 in funding in exchange for 6% equity. "Not only do food-oriented startups get access to mentorship from top minds in the food industry, but founders also receive support and resources to accelerate their businesses during the three-month program, all in preparation to make investor pitches," the ARK's Jeannette Balleza told Reuters.

WHAT ELSE DOES THE FOOD INDUSTRY BORROW FROM SILICON VALLEY?

According to GigaOm, "2013 is turning out to be the year when entrepreneurs and investors are embracing using technology to create more sustainable food products, from plant-based egg, meat, and cheese replacements, to healthier candy and salt." Beyond Meat—backed by San Francisco's Twitter founders Evan Williams and Biz Stone via their company Obvious Corp—created an eerily accurate chicken substitute, for example. Given Silicon Valley's 2013 penchant for food start-ups, in addition to Obvious Corp, Khosla Ventures, PayPal founder and venture capitalist Peter Thiel's Breakout Labs, SV Angel, Kleiner Perkins Caufield & Byers, True Ventures, celebrities from Hollywood (Matt Damon), pro football (Tom Brady) and the tech world more broadly (Bill Gates) have also joined in. Some investors say food-related start-ups fit into their sustainability portfolios, alongside solar energy or electric cars, because they aim to reduce the toll on the environment of producing animal products. For others, they fit alongside health investments like fitness devices and heart rate monitoring apps. Still others are eager to tackle a real-world problem, instead of building virtual farming games or figuring out ways to get people to click on ads.

In the last year, venture capital firms in the valley have funneled about $350 million into food projects, and investment deals in the sector were 37% higher than the previous year, according to a recent report by CB Insights, a venture capital database. In 2008, that figure was less than $50 million. That money is just a slice of the $30 billion that venture capitalists invest annually, but it is enough to help finance an array of food start-ups.

Venture capitalists have strayed from pure technology to food before. Restaurant chains like Starbucks, P. F. Chang's, Jamba Juice, and, more recently, The Melt, were backed by venture capital. Recipe apps and restaurant review sites like Yelp have long been popular. But this newest wave of start-ups is seeking to use technology to change the way people buy food, and in some cases to invent entirely new foods. Investors are also eager to profit from the movement toward eating fewer animal products and more organic food.

Khosla is backing San Francisco-based Hampton Creek Foods with their product "Beyond Eggs," who obsess over eggs and how these characteristics can be replaced with a combination of plants. The team has worked on over 344 prototypes for its egg-yolk product, and has studied 287 types of plants that range from peas and canola to make an egg substitute. Furthermore, they are working on egg-free mayonnaises, sauces, and dressings. Khosla Ventures also backed a few other food and agricultural companies, including artificial salt company Nu-Tek Salt and fake meat company Sand Hill Foods. Breakout Labs has invested in Modern Meadow, a company that aims to "print" lab-grown meat and leather. All of these companies are challenging the common nutrition advice to eat whole foods and vary your diet.

The next big challenge the food industry will tackle is food waste. Mintscraps—an online waste tracking platform for restaurants and supermarkets and an easy way to track waste, visualize savings, and divert or donate food scraps to food waste haulers or food banks—addressed this challenge well from a technical point of view, while Food Tank, the food think tank tackled it well from an educational point of view.

In addition to food incubators and innovations labs, the food industry is also borrowing "hackathons" from Silicon Valley and is launching "food hackathons." According to Wikipedia, a hackathon (also known as a hack day, hackfest, or codefest) is an event in which computer programmers and others in the field of software development, as well as graphic designers, interface designers, and project managers, collaborate intensively on software projects. Occasionally, there is a hardware component as well. Hackathons typically last between a day and a week in length. Some hackathons are intended simply for educational or social purposes, although in many cases the goal is to create usable software.

So what is a food hackathon? A food hackathon unites designers, developers, and entrepreneurs for a world class hackathon competition, to build networks, cross-pollinate ideas, and create new products and tools to innovate and improve the food ecosystem. Food Hackathon is the first event of its kind, empowering food lovers and developers, with a focus on building hardware and software products and services that positively impact the production, storage, distribution, access, discovery, sharing, consumption, and social impact of food. Top Food Hackathon teams will have an opportunity to demo their products and receive feedback from notable judges with a chance to win prizes and awards, gain recognition among the global food and tech community, and network with notable movers and shakers in Silicon Valley.

Coca Cola is also sponsoring a hackathon to re-imagine that most boring of technologies: the soda fountain. What would you do? Miniaturize the technology? Make one for the home, or for on an airplane so flight crews don't have to load and unload so many cans? The possibilities are huge! Projects can be submitted in any format from a business plan to a fully functioning prototype, but more completely proven ideas will be considered far easier to commercialize. The winning team gets their prototype and intellectual property related to the project purchased by The Coca-Cola Company at the end of the weekend for $10,000. If teams wish to keep their IP, they can decline the prize and somebody else is declared the winner.

In order to make it all happen and to get everybody up to speed on the carbonation process and other things related to making a fountain, The Coca-Cola Company taught a class, provided two demo units to take apart and tinker with, and teamed up with Hacker Dojo as partner and host. Hacker Dojo is a co-working space/community center for hackers and thinkers located in Mountain View, CA. Even though Hacker Dojo had limited physical resources, they provided a small CNC (Computer Numerical Control) mill and a 3D printer. Participants needed to bring their own metal for the mill and were kindly asked to chip in to help cover the cost of the extrusion for the 3D printer. Hacker Dojo also had a wide variety of electronics pieces in their electronics lab and in a nutshell, participants were ready to play. The idea was judged by a group of representatives from The Coca-Cola Company and a selected local "celebrity judge" from the start-up world.

More Examples

Meet Fast Food 2.0
Another disrupting example for food systems comes from former McDonald's executives Mike Roberts and Mike Donahue. People like fast food because they are cheap and taste yummy to many, but is there a future for sustainable fast foods? In October 2011, the two Mikes opened Lyfe Kitchen in Palo Alto with the idea of applying what they have learned in the fast food industry to this new culinary program. Lyfe aims to be a radically sustainable and healthy fast food chain with affordable prices.

Meet Nutrivise: Hipmunk for food
Hipmunk is a travel search site that aims to take the agony out of travel planning. It presents search results chronologically by time of day on a single page. Flights are ranked by price, schedule, and "agony," a score based on factors such as duration of the flight and number of stops. Nutrivise builds a personalized meal recommendation engine to help users reach their health and fitness goals efficiently. Their algorithm sorts through thousands of recipes, meal replacement bars, and chain-restaurant dishes to create daily meal plans that fulfill your needs and preferences whether you are eating at home, out, or on-the-go.

The Travelling Spoon: Eating the world, one spoon at a time
San Francisco-based Travelling Spoon connects travelers with authentic food experiences in people's (pre-vetted) homes around the world. Sounds a bit like San Francisco-based Airbnb.

Fooducate: empowers consumers to make better, healthier food choices
The Fooducate app is a top-ranked health app that gives you a nutrition grade plus bullet point explanations, by scanning the barcode of any food product, and offers you to select healthier alternatives. Similar to Fooducate, eBay's *RedLaser* barcode-scanning app let's you compare shopping and find product information on the go. It allows users to scan a barcode on an item at a store and then automatically access any eBay listings of the product on the marketplace. Sellers can also use the scanning technology to scan an item and list the product in very little time. Pivot is a word that tends to be over-used in the tech world, but in eBay's case that is exactly what we are witnessing with all their recent acquisitions—a major pivot in the company's business model to local commerce. With every change and disruption in the technology field, the nutrition, health and wellness, and retail industries get an opportunity to be part of it.

Postmates: the Uber of packages
As we learned in the other chapters, household dynamics are changing. *Postmates*, a revolutionary same-day urban logistics and delivery platform brings some disruption to the courier space once and for all.

Postmates is hoping to transform this niche industry in the same way that Uber is disrupting elite limo services and taxicabs with an on-demand delivery service (no more flagging or yelling required) that connects local couriers and bike messengers to anyone who needs to ship anything—from birthday cards to grand pianos—within a city in under one hour. In April 2012, Postmates launched their new Get It Now service in closed beta. Get It Now extends the Postmates delivery service with the ability to purchase goods on behalf of the user at any retail store or merchant in a city. The value proposition of Postmates is clearly that it aims to unlock a huge market of local commerce within a city, with the goal of having hundreds of local bricks-and-mortar businesses in a way that only flower shops have seemingly managed to do today.

The jury is still out how much Postmates will disrupt companies at the top of the food chain like FedEx and UPS (United Parcel Service), but also Safeway.com, Amazon.com or Amazon Fresh, Google Shopping Express, etc. Safeway Inc. is the second largest supermarket chain in North America, after The Kroger Co. Safeway.com is their e-commerce shopping platform offering a same-day delivery when you order before 8:30am, in addition to recipes and meal tips.

Chefler

Another start-up more in the prepared meal delivery business is San Francisco-based *Chefler*, which brings healthy meal plans for busy professionals to their home.

A FOOD REVOLUTION BEYOND SILICON VALLEY

A new coffee lid

Not all innovation comes from Silicon Valley and "The Valley" also admits this, of course.

It can also be *a new coffee lid from Kalamazoo, MI*, thanks to The Wall Street Journal highlighting it in February 2013. US lid demand, one study estimates, will show a 4.9% yearly rise to $1.2 billion in 2016. Mr. Bailey is not the first to tackle tops. The first patent for a drinking lid was issued in 1935. The archetypal Solo lid was featured in the 2004 Humble Masterpieces exhibition at New York's Museum of Modern Art, and a 2007 show at the Cincinnati Museum of Art featured a range of tops. But Mr. Bailey had his own ideas. He tinkered in his kitchen, using clear plastic prototypes and colored water to test liquid movement in the cup and spillage patterns.

He insisted on the triangular drinking hole and the slant through which to coax the coffee. He added the middle aerating hole to create fluid dynamics, ensuring calm waves of liquid throughout, rather than aggressive rushes up and over the sides. Because warm milk reacts with more-common polystyrene tops, making them brittle and loosening the seal over the cup's rim, he chose a more flexible material for his lids. You can use them more than once. FoamAroma lids are made at a penny apiece at a factory in Florida, not Taiwan. Starbucks chose the lid for its experimental new Tao tea store in Seattle. Amid all the beverage-industry hoopla, Mr. Bailey noticed what others had not: coffee makers and proprietors, behind the counter, sip from nice (top-free) ceramic cups. "Most baristas and coffee-shop owners do not drink from paper cups. They don't experience what customers do," he says, and so they miss the real point, "A lid is not just a lid."

Printed meal, anyone? A 3D Printer for Dieters?!

We have all heard of 3D printers that build plastic-like prototypes and even devices that 3D-print food. But here is Pablos Holman's nifty twist: a food printer intended to let people eat what are basically customized diets, while helping to eliminate food waste. Pablos Holman, an inventor at Intellectual Ventures, built a device that gives users exactly the amount of food they want, on demand. But instead of printing ink droplets, it deposits droplets of food. (Source: blog.machinedesign.com.)

3D Printing

It's something you hear talked about every day in the San Francisco Bay Area currently. The concept is simple: take some colorful plastic, heat it up until it melts, then extrude it with a precise machine to build any object, layer by layer. 3D printing has been around for more than 30 years. What has changed in the past months is that it is finally making its way into offices and homes with desktop-size printers that cost about as much as an average laptop. Want to prototype your own ideas? You can start playing with 3D design tools like 123D from Autodesk. The objects that you have been dreaming of can become tangible in a matter of minutes. Imagine, design, print, improve, and print again.

Habitat "smart world"

We live in a "smart" world. Cook and Das define a smart environment as "a small world where different kinds of smart device are continuously working to make inhabitants' lives more comfortable." Smart environments aim to satisfy the experience of individuals from every environment, by replacing the hazardous work, physical labor, and repetitive tasks with automated agents. A revolution in technology is allowing previously inanimate objects—from cars to trash cans to teapots—to talk back to us and even guide our behavior.

Meet HAPIfork, the smart fork

HAPILABS was founded in 2012. Its vision is to help individuals in the twenty-first century take control of their HAPIness, health and fitness through applications and connected devices. HAPIfork, the flagship product of HAPILABS was unveiled at the Consumer Electronics Show (CES) conference in January 2013. Studies have shown that the problem is not only *what* we eat but also *how fast* we eat. HAPIfork informs us if we are eating too fast. Like your smart toothbrush urges you to spend more time brushing your teeth, or smart sensors in our cars can tell if we are driving too fast or braking too suddenly, the HAPIfork is a smart fork that can monitor how quickly you are eating and signal you with indicator lights to slow down. Other products in development include the HAPIwatch to help people sleep better and the HAPItrack to help people stay in great shape.

Good smart versus bad smart: smart shopping with smarter carts

How can we avoid completely surrendering to the new technology? The key is learning to differentiate between the "good smart" and "bad smart." Devices that are "good smart" leave us in complete control of the situation and seek to enhance our decision-making by providing more information. For example: a grocery cart that can scan the bar codes of products we put into it, informing us of their nutritional benefits and country of origin, enhances—rather than impoverishes—our autonomy. A prototype has been developed by a group of designers at the Open University, UK.

Microsoft has also demoed a smarter cart equipped with a Mircorosft Kinect sensor bar, a Windows 8 tablet, a UPC scanner, and a motor. Think of it as your personal shopping assistant that follows you around the store, except in cart form. In other words, shoppers will be able to load the cart with their grocery list and the cart will in turn tell the customer where in the store the products are located. The cart also can help shoppers be health-conscious, asking whether or not the shopper may have wanted the gluten-free version of a product instead. Once finished, the shopper can actually skip the checkout lines and pay for their groceries right at the cart.

Technologies that are "bad smart," by contrast, make certain choices and behaviors impossible. The most worrisome smart-technology projects start from the assumption that designers know precisely how we should behave, so the only problem is finding the right incentive. The problem with many smart technologies is that their designers, in the quest to root out the imperfections of the human condition, seldom stop to ask how much frustration, failure, and regret is required for happiness and achievement to retain any meaning.

Smart kitchen

To grasp the intellectual poverty that awaits us in a smart world, look no further than recent blueprints for a "smart kitchen"—an odd but persistent goal of today's computer scientists, most recently in designs from the University of Washington and Kyoto Sangyo University in Japan. Once we step into this magic space, we are surrounded by video cameras that recognize whatever ingredients we hold in our hands. Tiny countertop robots inform us that, say, arugula does not go with boiled carrots or that lemon grass tastes awful with chocolate milk. This kitchen might be smart, but it is also a place where every mistake, every deviation from the master plan, is frowned upon. Rest assured that lasagna and sushi were not invented by a committee armed with formulas or with "big data" about recent consumer wants. Truly smart technologies will remind us that we are not mere automatons who assist big data in asking and answering questions.

Going back to the HAPIfork example, it is also going to Kickstarter for funding, like Forage Kitchen, as we learned earlier. With every day, Kickstarter gets populated with new food campaigns and culinary causes you as an individual can get behind and support, be it from a portable French press to a man who wants to resurrect the knish. A knish or knysh is an Eastern European and Jewish snack food made popular in North America by Eastern European immigrants. Another very popular crowd-funding tool is Indiegogo, which also hosts a plethora of food-related projects. Crowd-funding, as its name suggests, is a funding method where common people like you and me, henceforth the crowd, fund your personal or business project with their own money. The main difference between crowd-funding and donation is that crowd-funding is tied to the American JOBS Act that allows online sales of small stock to a huge pool of investors. Such tools like Indiegogo, Kickstarter, etc. have lowered the entry barrier for innovation from anywhere.

FUNDING GOOD DESIGN IS NOW OFFICIALLY MAINSTREAM

Good design, as defined by Dieter Rams, depicted in Figure 12.5, with special emphasis on the food industry was already quite extensively discussed in Chapter 1 (Universal and inclusive design). Let me add a few more general considerations. My examples in this chapter show a strong synergy between food and technology. More and more design-based start-ups are gaining traction. Start-up companies like Airbnb, Pinterest, and Etsy, among others, paved this path and show that they can also get money for their ventures.

The latest to push this trend forward is Kleiner Perkins—a firm that probably has the highest brand recognition among the VC crew to Valley outsiders, according to GigaOm—a Web 2.0 blog. At the end of 2012, Kleiner announced that it was adding a Design Fellows program, where design students would be able to apply for a three-month internship with

Good design is . . .

Figure 12.5 What makes a design good? Dieter Rams' 10 principles for good design. (Reproduced with permission of Wells Riley).

top companies in Kleiner Perkins' portfolio to hone their design skills. Design-oriented Kleiner portfolio companies, include start-ups like Nest, Flipboard, Opower, and Path.

What Kleiner Perkins is doing isn't totally novel. Google Ventures has been stockpiling design talent for its Design Studio, and is using that as a strategic edge to attract early stage deals. Dave McClure's start-up accelerator 500 Startups launched the Designer Fund, which allocates funds for design founders, in the spring of 2011. As the creator of the fund, Enrique Allen, explained in Fast Company, the fund is "about helping to give designers a seat at the start-up table that engineers and MBAs already have." 500 Startups also hosts its annual Warm Gun conference, which focuses on product design, and 500 Startups' portfolio companies commonly get schooled in previously unconventional skills like marketing and user interface design. Designer founders they have observed are consistently multi-disciplinary and have the cross-functional skills necessary to make decisions about products.

At the same time, design firms like Ideo are launching their own incubator startup programs. Rather than work for just the biggest companies in the world, they will be working with some of the smallest. It is part of their new Start-Up in Residence program.

Designers, start-ups and investors see the union as productive for all. The trend has become so pervasive that Fast Company even declared a design bubble brewing for venture capitalists, starting in 2011.

Being a designer founder of a tech start-up is one of the biggest opportunities in the design field today. What is driving this movement?:

1. As the consumer tech market becomes more crowded, brands and experienced design—not just technical capabilities—are becoming critical to success.
2. Long neglected in practice and academic scholarship, design patents (versus utility patents that focus on the function of a product) have exploded in importance as a result both of recent changes in the law, and high-profile cases like the Apple versus Samsung infringement case. Products and services from companies such as Nike, Apple, Samsung Electronics, and Google are not just about the function, but as much about the design.

The year 2012 saw the revival of the generally overlooked design patent, and the growing importance of trade dress. The increased recognition of these rights provides the entrepreneur additional opportunities and a burden to protect innovation. Design patents are a type of industrial design right granted for the ornamental design of a functional item. Ornamental designs of jewelry, furniture, beverage containers, consumer appliances, smart phones, USB devices, and computer icons are examples of objects that are covered by design patents. Perhaps the most notable patent litigation of 2012 involved design patents covering the iPhone, which Apple asserted against Samsung.

After decades of utility patents being the primary source of protection for innovation in the high tech industry, it is telling that the battle of the next decade began with a court's wide-reaching recognition of design patent rights. Trade dress protects consumers from the packaging and/or appearance of imitative goods and/or services, and is intended to prevent consumer confusion. Some examples of trade dress are the shape, color, and arrangement of the materials in a children's line of clothing, the design of a magazine cover, the appearance and décor of a chain of Mexican-style restaurants, and a method of displaying wine bottles in a wine shop. Importantly, in the past few years, many jurisdictions, including the USA have provided trade dress protection for the overall "look and feel" of a website.

3. Innovation is about radical collaboration. The critical mass of combined design, technical, and business skills enables product iteration to happen faster and at a higher resolution.

Designer founders have unique skills (not just visual) to understand human needs and discover unarticulated opportunities, but the trend of web and mobile start-ups using good design principles—and placing more emphasis on product development, UI (user interface), and UX (user experience) is just emerging. The Internet is maturing—regular people are immersed in using it now. Think of Pinterest—a pinboard-style photo sharing website that allows users to create and manage theme-based image collections such as events, interests, hobbies, and more. Users can browse other pinboards for inspiration, "re-pin" images to their own pinboards, or "like" photos—and there is an onslaught of app (application)-overload brewing, where good design will be the leading way to success.

The Internet and mobile apps are both a visually dominant and interactive medium, which means design and UI needs to be baked into the creation process from the beginning. Designers that make connected gadgets have long known this, as has Apple. Now start-ups like Instagram, Pintarest, Tumblr, and Path are using design as a differentiator. Instagram is an online photo-sharing and social networking service that enables its users to take pictures, apply digital filters to them, and share them on a variety of social networking services. A distinctive feature is that it confines photos to a square shape, similar to Kodak Instamatic and Polaroid images. Tumblr is a microblogging platform and social networking website. The service allows users to post multimedia and other content to a short-form blog. Users can follow other users' blogs, as well as make their blogs private. Path is the simple and private way to share life with close friends & family. And of course, where highly valued start-ups go, angels and VCs will always follow (Figure 12.6).

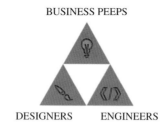

BUSINESS PEEPS

DESIGNERS   ENGINEERS

Figure 12.6 Innovation happens when designers, business people, engineers, ... come together. (Wells Riley, Startups, this is how design works, http://startupsthisishowdesignworks.com, 2013. Reproduced with permission from Wells Riley).

WHO ARE THE FOOD AND DESIGN START-UP PLAYERS?

Meet Good Eggs, the Etsy for Local Foodies
Etsy is an e-commerce website focused on handmade or vintage items, as well as art and craft supplies. Good Eggs is a San Francisco-based technology company that is driven by a mission to grow and sustain local food systems worldwide by building a one-stop online shop that handles the transactions and user interface for local food vendors. It is similar to Etsy in that Good Eggs gives small vendors an avenue to sell their products online without having to build their own site. For instance, it connects you to Fresh Baby Bites for delicious baby foods made fresh by Kim and Cristina or to Three Babes Bakeshop for pies where you are a mouse-click away from eating it in your house or feeding purees to your infant.

Foodspotting: finding and sharing great dishes, not just restaurants
San Francisco's Foodspotting Company, acquired in early 2013 by OpenTable, is a visual guide to good food and where to find it. Instead of reading and writing restaurant reviews, you can find and share great dishes. Browse photos of nearby dishes and see what looks good or follow friends and experts like Chow or Travel Channel to see what they love. Find restaurants and see what foodspotters and friends love there. The idea of foodspotting is to cover the earth with amazing food sightings, from childhood hometowns in Pennsylvania to food capitals like Tokyo.

Ness, short for likeness
Ness added more than 100,000 users and generated more than 1.5 million ratings in just a few months, a feat that took Yelp Inc., a local search and reviews online service, three years. Ness provides recommendations via "radical personalization," but also integrates with a user's social media Foursquare check-ins and Facebook friends' ratings in an unprecedented way. CEO Corey Reese has plans for taking Ness beyond food recommendations: "We want to build a digital replication of a person's preferences."

CONCLUSIONS

In this chapter we learned and analyzed the following areas:

- What a connector and a maven are
- Connectors in action and formal programs and approaches for scouting the right innovation partnerships

- What it takes to become a connector and to be aware of the surrounding soft facts of networking
- Silicon Valley and its unique eco-system:
  - Co-working spaces, incubators, and accelerators
  - Food and tech related co-working spaces, incubators, and accelerators
- What the food industry borrows from Silicon Valley and its trends
- More food and tech examples of the current food revolution outside of Silicon Valley
- Design and tech start-ups and its design bubble
- Design and tech start-ups in the food and retail industry.

If you look underneath the surface, you now recognize that the common denominator is connections—be it your network and/or someone else's network you choose to actively reside in.

TOPICS FOR FURTHER DISCUSSION

- Are you a connector or a maven, or do you prefer to rely on other people's networks?
- Do you have success stories where your network or somebody else's network played a key role?
- Who did you meet in your life that had an impact on your career, venture, projects, friendships, etc.?
- Were you part of an incubator or accelerator or community you can recommend to other fellow tech and design entrepreneurs in the nutrition, health, and wellness space?
- What were your experiences in your hometown where you built and launched a food and tech venture? What obstacles and benefits did you experience?
- What are your top three favorite tech and design start-ups in the food industry?

13 What was it all about? An attempt at a conclusion

I wear glasses with yellow lenses because they make everything look brighter.

Leonard Cohen, private conversations, 1970

A FEW MOMENTS IN THE LIFE OF MANNY MIDDLE

Manny Middle is an employee in a large food company. I could also have told the story of Minnie Middleton, but let's stick to Manny, at least for now. Manny has a responsible job, he is a department head and responsible for a fairly large group of people; they work in product development and Manny is very proud of what he is doing and what he and his group have achieved, at least as far as he can judge. Much of what they do is funneled towards what they call "conceptualization," another word for "let's try it out, if it doesn't work, no harm done." Manny has a boss, who has set a number of great objectives for him, and Manny really tries hard to achieve these. So, in turn, he sets all his people a number of great objectives, and at the end of the year they all will be happy to have achieved these, or maybe even over-achieved; what a feast!

His boss has an obligation to not only push everyone to come up with successful concepts, which eventually will become successful products in the marketplace, but he is pushed by his boss to do all this at the lowest costs possible, hence the degree of achievement is not necessarily seen the same way by Manny and his boss. Remember "the costs, stupid!" Manny is squeezed in the middle of all of this, he is the proverbial "middle manager." He tries hard, but often this is not good enough and he is doing what many smart people do: first, he adapts and second he tries to understand by learning even more about his job, the company, and the industry at large. The learning is even proposed by his boss: "Take this course, the company offers it, and you will learn everything you need to know to become a better manager."

Food Industry Design, Technology and Innovation, First Edition.
Helmut Traitler, Birgit Coleman and Karen Hofmann.
© 2015 John Wiley & Sons, Inc. Published 2015 by John Wiley & Sons, Inc.

On top of this, it takes you out of the daily treadmill, and you are going to meet like-minded people who are in the same situation. Best of all, the company pays for it, even for your meals and coffee breaks, and all the other items such as hotel and even the evenings at the bar; after all, it's part of the experience: networking! And all this for a whole week, just imagine. And maybe, by doing this, Manny will not only come out as a better manager, better understanding how to deal with being squeezed, but he may even qualify in the not too distant future to be promoted and gain a level of responsibility where there is a different kind of squeezing going on. And maybe there is more to come: Manny might be tapped to go to one of the executive management courses, lasting several weeks and led by the saints in the consultancy industry, the teachers of the holy grail of coaching. Wow, that would be something. Yes, there are long days with late evening work, in important-looking working groups, with highly ambitious topics, for which some of the "wanting to be even smarter" participants have already brought pre-fabricated slide presentations, pretending they have done these between dinner and coffee.

Best of all: they cost a lot of money, so they must be great, and Manny will be one of the selected ones, once he comes out of such a course, which may cost the equivalent of one year's salary for a lab technician; wow! Remember, Manny is still in his first one-week management course, and most likely, and without any cynicism, comes out a better person. He has even realized that he knows less than before; let me re-phrase this: he didn't know how much he didn't know when he enrolled in the course. Now, Manny starts his own crusade and he remembers those bookstores at airports when he goes on a business trip, which are packed with overflowing bookshelves stuffed with so-called management and business literature. He starts buying these, but not before he has asked his boss what he would recommend. So, Manny buys and buys, and starts reading these books. Let me emphasize: he "starts" reading these books, until he realizes that these are not novels and there is no plot. Manny, by training is a linear thinker and there is no real linearity in these books. How does he know what he should read and how he should read and digest what he has read? Panic sets in but, Manny being Manny, namely an expert in being squeezed, he will make it out of this situation with bravado, and he will learn with every book that he dissects and to which he applies the technique of "lectio selectivam" (selective reading), he becomes better at this, and the devouring begins. And there is one more important lesson that comes with it: Manny has become innovative and this out of sheer need! Being pushed and squeezed is a great trigger for finding innovative solutions.

Manny starts to understand, not only his own area of squeeze, how all this comes together and where his industry begins, how it works, and where it is supposed to end and re-start again.Let's go on a short trip with Manny; it will bring this whole book together in a fun and hopefully entertaining way, like a comic book. This is a real field trip, namely from "Field to Fork" and transits through all of the important stations that have been discussed in this book, like design, technology and innovation, and ultimately the consumer.

FROM FIELD TO FORK

(All illustrations by Pierpaolo Pugnale, aka "pecub", summer 2013.)

Harvesting the right raw materials is crucial for the food industry and is meticulously done and closely inspected. 	Ever so often, there are new agricultural techniques developed and applied, which may even yield new designer crops. 
Innovation and new technologies more recently led to alternative agricultural practices. 	The consumer is at the beginning and at the end of all that is done in the industry. Here, consumer expectations meet the technology push. 

Nutrition has become the driving force in product development; medical food is the new nutrition.

Food safety includes traceability, which really begins in the field.

The search for new formulations in foods and beverages has become very sophisticated.

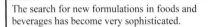

Performance food is the name of the game.

Once successfully developed, the products need to be manufactured on efficiently designed lines.

Production has become a new paradise, full of seductions and possibilities.

The consumer is discovered and defined in new ways. It was found that the gut is the second brain, maybe even the first!

Products have to be tested and rated. Often, outstanding results are found through this approach.

Sometimes, management does not have any new ideas and calls for "back to basics!"

All of a sudden, new packaging design is required.

It is well known that great design creates outstanding packaging and food products with additional value.

And all this new and innovative packaging, which is perfectly in sync with nature.

The logistics manager has to manage people in outrageously complex projects.

Innovation has become open, mysterious, almost witchcraft with too many magicians.

Design sells: designer food – the eye eats too.

Food has become available wherever and whenever.

The profit margin race continues: EBIT, EBITA, EBIDT, EBIDTA, OP1, you name it, you get it.

Taste goes global: designer foods conquer the world, at least the part that can afford it.

Pets have become our most popular companions. The industry works hard to improve palatability.

Shopping is increasingly a fine balance between real needs, pleasure and nightmares.

We wish "Little Red Riding Hood" good luck on her way to the supermarket: choose wisely!

I do hope that you enjoyed the story and the excursion into the world of an illustrator, and how he sees the food industry and everything (or at least most of it) that is linked to it: agriculture and raw materials, supply chain, consumer pull versus technology push, product development, manufacturing, design, innovation that has become open, healthy and nutritious foods and beverages, globalization of taste, the importance of pet care products, and, last but not least, the ultimate shopping experience.

MANNY MIDDLE DISCOVERS THE ROLE OF DESIGN

Let me come back to Manny, the "new" Manny, the one that has started to read all about his industry, how it works and how it "ticks," and who begins to understand how many elements, especially innovation and technology, play a crucial if not a central role in this industry. Manny wants to talk to other colleagues outside his direct area of work. He talks to his boss and is encouraged to start networking internally, meet with people from marketing or finance, supply chain or procurement, operations or human resources, sales or legal, design or communications, public relations or consumer affairs; in other words, Manny should start to discover and, to some degree, understand the "value maze"; not an easy undertaking, and this promises to be an exciting journey of discovery.

This is how Manny meets Minnie Middleton; she works in procurement. When she tells him about her job, he only hears costs and money and contracts and term sheets and negotiations and timelines, transport problems, delivery requirements, preferred suppliers and volume-dependent pricing, supplier-managed inventories, reverse auctions, and, and, and … All this sounds very unfamiliar and Manny is highly confused. He and his team are supposed to invent and develop great, new, and successful products for the future wellbeing of the company, and Minnie seems to put a question mark on all of this. How could she do this and how can the two worlds, the world of high innovation and the world of low costs, be brought together or at least closer to each other? It's not an easy task because the drivers are so different. What's even more different are the objectives that Manny and Minnie have been set, and which they try to achieve for their own benefit, but also for the benefit of the company at large. These objectives seem to be so different, so far apart, and sometimes even contradictory.

There is no single recipe to successfully bring these two worlds closer to each other, and it involves a tremendous amount of commitment and hard work, spiced up with humility and a strong desire to understand the other world. And it requires time that often is just not available, unless there is a very strong element of long-term thinking and durable sustainability in the company. The good news is, and this is totally based on long-standing personal experience, every single person in a company can influence the degree of long-term thinking in his or her own work environment. I once heard an interview with an apparently famous Japanese financial guru and investor, whose name I have forgotten, and who was asked how he runs his business. He answered in the following manner: "I have short-term, medium-term and long-term objectives." The interviewer then asked him to explain this in a bit more detail. "Well, short term typically is five seconds, medium term is about one to two minutes and long term is as long as 10 to 15 minutes." The point here is that the definition of short, medium or long term strongly depends on the type of industry, the people involved, the business model and the short-, medium-, or long-term financial expectations of your company, as well as the expectations of the financial analysts. I do not want to discuss or even try to analyze the role of financial analysts in the world of food companies. They are just there, they do a job that is apparently important for the benefit of the investors and shareholders, and that's that.

If a company wants to be truly outstanding and deliver excellence through great products and services, the role of financial analysts is as obsolete as the need for patents in an environment that works so fast that the real protection is speed to market and consistently being the first and most successful. However, neither of these two situations is realistic, hence the need for both, analysts as well as patents. And, as you know, both need each other: patents need the analysts to justify their existence, at least in part; analysts—in addition to other elements—need a number of patents from a company for their analysis. My point here is to show the complexity of the value maze, especially when external considerations such as satisfying the "market" come into play. Don't get confused by this, don't let yourself be distracted by all this, and pursue your quest for greater understanding, very much what Manny has set out to do.

Minnie and Manny have good discussions and decide to learn and discover more. They are especially intrigued by one fact: packaging in the company is mostly seen as a nuisance, it costs too much and it's just material that should actually be reduced. Minnie knows that their company spends about 15% of the annual revenue on packaging materials and packaging-related activities, that's more than a handful, that's an awful lot. And yes, one of Minnie's tasks is to reduce or "optimize" this spending. Together they discover one totally underestimated element, namely design, especially good packaging design. They decide to dig deeper into this question and find out all about the role of design in a food company. They are surprised that design plays such a discrete role. Typically, design is used to make stylish corporate presentations or splash one more color on an otherwise boring package.

They quickly understand that design could play a much more important and pivotal role in the food industry. After all, the packaging is the first interface that every consumer experiences when purchasing a food product, so one would think that much emphasis is put on the design of the package. And yet, cost optimization seems to be of a higher priority than great package and food design. They start to understand that something should change, and should change rather quickly so that design receives the level of attention that it has to have in the industry, especially in the food industry. Manny and Minnie wish for more courageous leaders who would spearhead such a change, but they also understand that they have a role too; they have to push this idea in their own work environments. Manny decides that he will involve industrial design in his next development project, especially for great packaging. Minnie plans a visit to a design school in town at the occasion of their upcoming open day event.

Minnie has had the opportunity to spend an entire day at the design school and has become fascinated to learn about the expanding role of design in many industries. As the industry shifts, design expands to new grounds, speed to market is totally supported by design, co-creation is the name of the game, and design serves as facilitator. She hears about the trend for going out of the design studio and into the world, how designers are valued today, leading towards a new mindset; she learns about the transformational impact that design can make, and about design as a new process of invention and a true catalyst for change. She also listens curiously to designers explaining the future of meaningful product experiences, delivering such experiences, and, last but not least, how design enables collaborations and empowers the leadership of innovators.

At the end of the day, she has become convinced about the role of design and what it could move in her company. She meets Manny and talks to him about her learning from that day. They realize that design could be a game-changer in their company. Everyone in the organization has a role to play and they decide that they will do their share. And she strongly believes that it would be great if their company had a "CDO" (chief design officer), someone who really could bring non-linear design thinking to his colleagues on the executive board of the company.

MINNIE MIDDLETON TAKES A CLOSER LOOK AT THE ROLE OF TECHNOLOGY AND DISCOVERS THE CONCEPT OF SUPPLIER PARTNERS

Procurement is a very diversified work area. It's never really boring because nothing ever seems to be the same and so many things have to be considered. Procurement, depending on the level of maturity of the company and the specific function, more or less relies on either next-generation IT support, increasingly in the cloud for the more mature, while the less mature ones still heavily rely on well-proven legacy systems.

Minnie knows this well, as in her company both approaches exist and it's not always easy to bring these together in sensible and successful ways. Moreover, there are so many different suppliers and many of them, simply for reasons of sheer size also rely on their own systems. What a mess, she thinks, and couldn't we be moving to more collaborative solutions?

She knows that improved analytics in procurement, combined with broader data integration (beyond one supplier and the company!) and collaborative tools (using the same IT systems) should lead to much improved efficiency. Procurement works with so many technologies, such as good old supplier relationship management, low-cost country sourcing (with all its safety concerns), inventory management, IT technologies in general, e-procurement vendors, e-auctions, the much hated reverse auctions (hated by the suppliers), value-driven sourcing decisions and strategic sourcing, and probably quite a few more. Despite all these, or maybe because, the personal element has got a bit lost. On the one hand this could be a positive as it is probably easier to negotiate when you leave personal feelings out. But Minnie believes that there still is a need for a personal element in procurement: it's the switch from a simple vendor–customer to a partner–partner relationship.

Minnie has heard that innovation partnerships are a great way towards such a relationship and she decides to add such a partnership approach as a new technology (or rather technique) to the already long list of technologies described above. When she studies and analyzes the partnership approach, she discovers that in this mode, she seems to receive a lot more back from supplier partners compared to the traditional and conventional modes of dealing with suppliers or vendors, as they are also called. She decides to find out more about this partnership approach, and asks her colleague Manny, who works in product development and is an important user of the outcomes of Minnie's negotiations, to help her in this endeavor. Manny knows that by working with partners he not only gets what he needs in a much more bespoke way, but he also knows that the partner puts more resources behind joint activities with their partner customer, Manny's and Minnie's company. They quickly realize the benefits and decide to not only continue to work in this fashion, but, more importantly, spread the word in their respective departments and beyond in the entire company. Let's hope that people listen to them!

MINNIE MIDDLETON DISCOVERS THE VALUE OF INNOVATION IN HER COMPANY

Now that Minnie has met Manny and they have been able to discover the potential of design and the importance of technology, she has more appetite to discover other important areas that she has not been fully exposed to yet in her present procurement role: innovation. Although many procurement activities heavily rely on innovative solutions, these are almost exclusively brought in from the outside, third party providers of solutions, with little or no role for those in procurement to be innovative or the creator of innovation at all. Manny works in product development and should actually know much better than anyone else in the company how to be innovative and create innovative solutions. She decides that Manny should be her teacher and tell her all about innovation and the role it can and must play in their company.

Manny talks about the essential stages that lead to value creation in their company, and explains that before innovation there is actually creativity. The desire to be creative, to become a creative thinker and actor, is a basic requirement for innovation of any kind to happen. This is not only valid for the area of product development, but is true for all areas of the company, from consumer research and marketing to manufacturing and sales, from human resources and communication to finance and legal, from advertising and business units to quality and safety and supply chain; in other words, everyone in the company, irrespective of where they work, has to think and act creatively and innovatively.

As a result of this thinking and acting one can proceed to the next stages in this process, namely creation of viable concepts that may lead to successful execution of these concepts and ultimately to value creation. By having contributed not only to great new products, but also to great new solutions of how to get there, all using the necessary steps, everyone in the company can contribute in innovative ways. Minnie is so excited about this realization that she sets up a meeting with her boss: she wants to propose the creation of a group within their procurement organization that specializes in "innovation in procurement" and that actively looks into creating their own innovative ways for all available IT systems and other tools to be introduced and fully integrated in well-thought through and original ways. Manny helps her in preparing this important event and coaches her by emphasizing the many facets of innovation, not only as one of the important stages of value creation, but also explaining and discussing the role of innovation as a connector between design and technology, and, last but not least, the role of innovators in the company.

They also discuss the various tools that innovators consistently and successfully use. It's a whole new world for Minnie, and Manny seems to get carried away with explaining creative problem-solving, the power of brainstorming, the divergence–convergence paradigm, thinking outside the box, thinking inside the box, the importance of fast and proper selection of winning concepts, and how to drive such concepts through every step in the organization to a successful execution and launch in the marketplace. Everyone in the company can use such tools for any purpose, appropriately adapted. Minnie is so impressed that she promises to spread the word in her procurement organization and beyond.

MINNIE MIDDLETON AND MANNY MIDDLE DISCOVER THE POWER OF NETWORKING AND TRAVEL TO SILICON VALLEY

Both Minnie and Manny know social networks, as each of them is subscribed to at least one of the more popular private, as well as professional, social networks. They have experienced themselves that networking can be great, but can be very tricky, if not downright dangerous, at the same time when not properly used and kept in check. Networking may even become embarrassing and sometimes they ask themselves the question "whom does it serve?" You may have experience in social networking too and might have had your own experiences, good or bad, or both. It is without doubt that, appropriately used, networking has the power to multiply forces and get things done, faster and more efficiently than without them.

They have heard about the role and power of professional networking in many areas of the world but none is better known and probably has brought more successes in innovation, invention, creation, and rewarding execution than Silicon Valley, a part of the San Francisco Bay Area in central California. They are curious and want to experience first-hand, so they decide to take a few days off and travel to Silicon Valley, and, through a friend and colleague, who not only lives and works there, but is extremely well connected and networked, they are confident that they will meet the right people to talk to and go to the right events to learn about great networking and professional connectivity.

Hardly have they arrived and met their friend, when they are introduced to several venture capital investors and learn all about what it takes to be successful in this business, be it as an investor or an investee. They hear and learn about Singularity University, corporate venture groups, and innovation labs, they spend time discussing bridge organizations, meet with a professional connector and connections explorer, discuss the power of networking, and learn some basic networking principles. They meet with more people in Silicon Valley and slowly start to get some insight into the Valley's ecosystem. They learn about food and tech incubators, accelerators, and how Silicon Valley is becoming bigger by creating co-working spaces in the greater San Francisco Bay Area, and across the country and beyond.

They are totally surprised to hear about initiatives in which the food industry borrows from the Silicon Valley approach and meet with "Nutrivise," "Fooducate," and "Postmate". They totally unexpectedly discover a kind of food revolution in and beyond Silicon Valley and learn about printed meals, 3D printers for dieters, Habitat "smart world," HAPIfork—the smart fork, good smart versus bad smart, smart shopping with smarter carts, and especially smart kitchen. They also realize that funding good design in all possible areas has now officially become mainstream and meet with several food and design start-up representatives like Good Eggs, Foodspotting, and Ness, short for Likeness.

What a great and exciting experience and the amount of learning in such a creative and innovative environment was unbelievable. But they have done it, they have got through, and cannot wait to get home, back to the workplace, and tell their colleagues, and encourage them to learn as much as possible about the great excitement that reigns around the food industry, even and especially in a tech-driven environment such as Silicon Valley and the greater San Francisco Bay Area. It is truly contagious!

Minnie Middleton and Manny Middle have finished their travels through the discovery of design, technology, and innovation in the food industry space and are convinced, more than ever, that all what they have read, heard, discussed, seen, and lived will make this industry so much stronger, give it a much higher value, and will increasingly lead to sustainable ways of making food that tastes good, is safe, and is affordable for the many people that this food industry serves around the globe.

Or, like my good friend Mike Cavaletto would say:

"So long, until we eat again"!

EPILOGUE: THE QUESTIONNAIRE

I have prepared a 10-point questionnaire regarding the role of design, technology, and innovation in the food industry, and have sent it out to a number of people in several food companies. So let me end this book by giving you an overview of the kinds of comments and answers that I have received from experts in the industry, from people who live design, technology, and innovation in the food industry on a daily basis, with all the joys as well as the frustrations that come with it. The first set of comments comes from Unilever and represents a top industry view given by their CEO Paul Polman. The second set of comments and answers is a summary of feedback from other industry representatives. As a reader, you may want to take the questionnaire as an assignment to meet with your peers and colleagues to discuss the individual points and find answers and comments that have relevance to your own work environment.

Figure 13.1 The new interest in food in Silicon Valley. (Reproduced with permission of Pierpaolo Pugnale, aka "pecub").

Feedback from the Unilever CEO

1. **What does "DESIGN" represent to you? An emotional response.**
   Design is about creating products that people want to buy or use over and over again. For food, that is often driven by the first time you use a product. So for example, the way that the chocolate breaks the first time you bite into a Magnum®, the crispiness of a Cornetto® cone. The way a Knorr® jelly bouillon has been created so that it looks and tastes exactly the way a chef would create a jelly stock cube from scratch. It's about creating food that tastes great and looks great.

2. **How can design have important implications in the fast-moving consumer goods (FMCG) industry, and, if knowledgeable, in the food industry?**
   Our business is one which relies on two billion consumers choosing to use our products every day. Therefore how we design our products at all levels is critically important. How the product looks on shelf, how the packaging works, and whether it is easy to use; does the product truly deliver against the claims it makes. All this is vital.

   But also the way we design for manufacturing. For example, we sell over 40 billion sachets of products each and every year to low-income consumers, so not only is it important that the product and pack is designed to exceed consumer expectations, but it also needs to be manufactured efficiently, and here, the pack design is critically important.

3. **Do you have personal and consumer-related experience with innovation in the FMCG industry? Can you give us some examples of successful innovations?**
   My career has been one of helping to drive a consumer- and customer-focused culture throughout the organizations in which I have worked. In the foods business and the FMCG industry we need to remember that the consumer always has a choice in terms of the products they buy. We have to serve those consumer needs better than the competition. This drives innovation.

   A good example would be Knorr Jelly Bouillon®. Bouillon is an important product with which to make stocks and soups, but it historically has been a dried product. When chefs make bouillon from scratch, by boiling down the stock, they create a jelly. Our technologists were able to recreate this process, and we now have a jelly bouillon, which is far superior to the competition, and consumers see a better product in terms of dissolution, flavor, aroma, and naturalness. It's been a fantastic innovation and contributed to a bouillon business with a turnover today in excess of €1 billion.

4. **How important would you see "TECHNOLOGY" in the food or FMCG industry? Would you like to see more of it or does it frighten or annoy you?**
   While Foods is perhaps less technology intensive than other parts of the FMCG industry, such as Personal Care or Home Care, it is nonetheless very important. I gave the example of Jelly Bouillon®. In our Spreads business, we developed a novel process to blend oils and fats, which means we now have products which are lower in fat and saturated fat, but give the taste and texture demanded by consumers. Again, an example of consumer-relevant innovation.

   There are of course aspects to technology beyond food technology. How we understand human nutrition for example. One of our biggest brands Becel/Flora® was developed in the 1950s as a response to what we see as a trend towards an increase in heart disease, so a product that lowered cholesterol (Blood Cholesterol Lowering BeCeL®) was developed. Areas such as nutrition, nutrigenomics, high-throughput screening will all play an increasing role in food technology in the future.

5. **Which roles could designers and design play in the FMCG industry?**
   - **Designing objects of any kind?**
   - **Assisting in product development?**
   - **Assisting in consumer and market research?**
   - **Consulting to marketing?**
   - **Co-decision-maker at board level?**
   - **Being a member of the executive board?**
   - **Anything else?**
   - **All of the above?**

   We think and talk about holistic brand design, so design thinking runs throughout the whole of our marketing and R&D organizations. Our Category Presidents are involved in the design of our brands; our Head of R&D is involved in the design of our products and our packaging. We have something called "Crafting Brands for Life" in our organization—the purpose of which is to truly understand the essence of a brand, and to ensure all those elements come alive whenever the brand touches the consumer. This includes every aspect of brand design, product, packaging, and advertising. So design thinking is represented at the most senior levels in the company and runs throughout the organization.

   In Foods in particular, we also emphasize the role of "chefmanship," which is design thinking brought to the world of food. We have over 100 chefs in the company whose role it is to craft our products, but also to ensure they work as part of a broader food menu.

6. **How do you personally support and encourage innovation in your role or function? Can you give one or two specific examples?**

   We have made "winning with brands and innovations" one of our key strategic thrusts and as CEO I have regular reviews of our long-term innovation programs with the combined Brand and R&D organizations. On an annual basis we review the delivery of that program against its objectives and plan the 1–3 year detailed project plans with the Categories and the business. It is a highly iterative and dynamic process.

   I have regular reviews with our Country Chairmen, and during those reviews, we always review innovation performance.

7. **Do you take a personal interest in supporting and coaching innovators in your work environment?**

   Of course. Innovation is the lifeblood of our business—so encouraging innovative thinking and nurturing a strong group of innovators within our Brand and R&D organizations is critically important. I review our key innovations on a regular basis and spend time in each of our main R&D laboratories around the world.

8. **How would you define the optimal and most appropriate way of compensating/coaching and mentoring innovators, be it individuals or teams? Can you give one or two specific suggestions, which, in your experience have worked or can work?**

   We have a very simple way of rewarding people, which is about setting stretching goals and then challenging everyone to achieve those goals. In areas such as innovation, that may be linked to the launch of a successful new product. In R&D,

it is linked to the value of our technology pipeline—do we have sufficient new ideas in our funnel to create the value we need from innovation over the next 3–5 years.

Mentoring is simply about creating a personal connection with key individuals and creating the space for reflection and conversations.

9. **How much time do you personally spend in coaching and mentoring talent?**
As CEO, at one level you could argue that my most important job is coaching and mentoring. After all, ours is a business of people and brands. I am travelling typically 50–60% of my time around the company, so in many ways every conversation I have is a coaching conversation.

The fact is we have put a huge emphasis over recent years in helping to build and develop our people, and especially our leaders. More than 500 have now been through what we call the Unilever Leadership Development Program (ULDP), an industry-leading program that puts the emphasis on authentic and purpose-driven leadership. I have been involved directly in the design of the program and in attending many of the individual sessions. I also have reviewed personally the development plans of all of our top leaders who have been through the ULDP.

All this of course in addition to the formal talent reviews that we hold at different points throughout the year.

10. **How do you see the role of design and designers evolving in the FMCG industry, and, if knowledgeable, especially in the food industry?**
Given I want design thinking to be at the core of the business, the role of design will continue to be very important to our Foods business. We need to create new innovations and new ways of providing benefits to consumers and that can only be done through designing better products and packs.

It is a process that is constantly evolving. Not only do we need to ensure we have the most talented people available working on this inside the company, but that we also continually evolve to work with the best external designers and—in the case of foods—with leading chefs as well, so that we truly capture the best design thinking that is available.

Other feedback, combined from several "voices" from the industry

1. **What does "DESIGN" represent to you? An emotional response.**
   - Design represents, together with functionality, the attractiveness of a product. Design may lead to an emotional bonding and is able to generate so-called "wow" effects. People come up with comments like "cool," "impressive," "remarkable," which shows that the corresponding product sticks out of a group of products with average properties. In the end design is often the deeper reason for a closer relationship of a consumer with the product and with the corresponding branding. In many cases selected elements of branding are adopted by the consumers.
   - *Aesthetic differentiation.*
   - A way of skillfully combining function to "do the job well" with form, "how using it makes me feel".
   - *For me, design is the way to communicate and provide innovation to consumers. Moreover, it is a very tangible way to create a positive relationship between a product or service and a consumer. To date, design is a paramount and absolute*

necessity in any leading product in the marketplace. How to create innovations that consumers do not expect, but that they eventually will love. How to create products and services that are so distinct from those that dominate the market and inevitably will make people feel passionate about them.

- Design has multiple facets, so first of all let's clarify that I see at least three different areas in design: graphic/communication design, industrial design, and its integration within the business, i.e. design thinking.

  Industrial design takes abstract ideas and turns them into tangible products to commercially exploit them at a mass scale through the use of a combination of applied art and applied science to improve the aesthetics, ergonomics, and functionality, the aim being to deliver superior user experiences.

  Communication design is the process of choosing and organizing words, images, and messages into a form that communicates and influences its audience. It uses different formats like posters, websites, and packaging graphics.

  Design thinking is a mindset and a set of methods integrated into our business in order to help identify real user needs and opportunities that will drive value.

- *DESIGN is important for me because all food products (services and systems too) are subject to the C.H.E.F.S. principles: Convenience, Health, Emotions, Functionalities, and Safety.*

2. **How can design have important implications in the fast-moving consumer goods (FMCG) industry, and, if knowledgeable, in the food industry?**

- In the food industry criteria of differentiation at the point of sales are of high importance. Different consumer groups reflect on different product features like visual appearance, aroma, taste, easy to use functionality, and overall quality. At the same time consumers do not only buy a food or beverage product. Consumers also search for innovative ideas and in particular also for new aroma and taste sensations like in confectionary products, for example. In beverages the whole area of non-caloric sweeteners is subject to disruptive innovation. New food additives, like Rebaudioside A from Stevia leaves extract, are marketed as a natural ingredient. Consumers are open for materials from natural sources and enjoy the effect of low calorie content. At the same time, beverage producers are aiming at cost reduction scenarios based on the reduction or omission of sucrose or high-fructose corn syrup, which will lead to a totally different portfolio of beverage brand items mid-term. In this context, design helps to transfer the message of green and natural ingredients to the consumer, for example in Argentina with Coca Cola Life®, with a green label and a combination of Rebausioside A and sucrose to provide authentic, natural sweetness to the beverage.

- *Packaging innovation for differentiation.*

- Helping the industry routinely build into our products consumer usefulness and ease of use through simplicity, robustness and minimum material usage throughout the supply chain: plan—source—make—deliver—recycle.

- *Design-driven innovation creates a unique leadership position in the marketplace. In the food industry, unique features become the product characteristics, e.g. Nespresso®, M&Ms®, Coca Cola® bottles, Coors Light® beer can, etc.*

- Most performing businesses in the market today are integrating design in order to support long-term innovation and create competitive advantage, in my company, we are using design to:

(a) make the abstract tangible: designers are trained to live in a very abstract world and think creatively, but they are also trained to commercially exploit that creativity and turn it into something that works and is commercially viable; (b) design experiences not products: design is about creating experiences that go beyond just a product; design considers all the touch points in an experience in order to design something that connects emotionally on many levels.

Designers think differently: they think beyond two dimensions and tackle problems from multiple vantage points, bringing a different perspective to the business.

- *Design crystallizes pain, needs, and desires of consumers and gives them a voice and a face in the product development process.*

3. **Do you have personal and consumer-related experience with innovation in the FMCG industry? Can you give us some examples of successful innovations?**
    - Personal consumer-related experience with innovation in the FMCG industry is very much dependent on personal consumption behavior. In most cases product introduction from the IT and computer industry side like, for example, computer apps, cloud technology, digital navigation, digital photography, or touch-screen or tablet PCs are at the top of people's minds. In the food and beverages industry there are less prominent, but still remarkable innovation trends. From my perspective the introduction of taste-improved alcohol-free beers, cocoa- or polyphenol-enriched chocolates, premium taste coffee products, chewing gum products with teeth whitening and gum healing properties, or new snacks omitting frying or baking processes are product innovations which are addressing customer needs and expectations.
    - *No personal experience.*
    - No personal experience.
    - *Indeed, lots of experience with successes in product areas such as long-lasting taste chewing gum, probiotic yogurt and fruits, exotic fruit juices, low-acid humus spread, veggie patty etc.*
    - I've been driving industrial design and innovation for four years at Roche, eight years at Novartis and most recently one and a half years in my present company. I have personally contributed to introducing many innovations onto the market, through my personal work, or through collaboration with agencies worldwide. My best success lies in the introduction of "thin strips" medication for the OTC (over the counter) market in the USA.
    - *I have worked for 20 years in research and innovation in the FMCG, where innovations are few, but small-step improvements on brands are the main driver of growth. One of the best examples is the functionalization of coffee–creamer mixes with powders encapsulating gases for foaming. This innovation has combined a novel high-tech solution with an enlarged design space for product development of the brand, satifying consumer preferences and production cost reduction (the holy grail). It had been a "cash-cow" from its first day onwards.*

4. **How important would you see "TECHNOLOGY" in the food industry or FMCG industry? Would you like to see more of it or does it frighten or annoy you?**
    - The development of new and innovative technologies for the production of food and beverages is tightly connected with the change in the availability and price of raw materials and the energy consumption during processing. Besides the frying of foodstuffs, freeze-drying is one of the most energy-consuming technologies in

food processing. Recent developments indicate that for fruits, a combined technology with expansion of the food matrix via microwave and subsequent air drying might lead to a taste-improved product without the expensive sublimation of water step. This approach considers modern aspects of sustainable food production, helps to protect the environment, and shows potential for even further improvements. All new technologies which meet modern criteria of sustainability, like caring for nature, resources, and society are finding more and more acceptance with consumers and allow the new positioning of products with the support of new designs.

- *Very important. Increasing opportunities to convey health benefits via more nutritious food, e.g. soy beans with no trans fats; also creation of more affordable food via technology; would like to see more of it.*
- Extremely important; would like to see more well thought-through and focused technology applications.
- *Technology is extremely important and provides the first and foremost advantage and the platform for delivering value to the consumers. However, it is invisible to the consumers. Consumers are quite afraid of technology, particularly when it comes to food, and the less they know, the better the product is perceived. This "phobia" has vast ramifications on product acceptance. Technology provides the marketing edge, but should be kept in the background.*
- Science and technology is a key driver for innovation, and designers are tailored to connect consumer insight with scientific/technical solutions. I see technology as an enabler and I am not afraid or annoyed by any of it. As a designer, I believe in our ability to identify, develop and use technology to the benefit of the entire community.
- *TECHNOLOGY, all too visible, is repellent for the consumer. The Third Law of Arthur Clarke is also valid for FMCG's products: any sufficiently advanced technology is indistinguishable from magic.*

5. **Which roles could designers and design play in the FMCG industry?**
   - **Designing objects of any kind?**
     The role and value of designers in the FMCG industry is very similar to the role and function of flavor specialists in the flavor industry. Both are the interface between product properties and consumers, both are using human creativity and emotional intelligence, as well as passion and patience to create a new impression out of details. This capability can be used for the design of packaging, for the design of taste and visuals and for the creation of new experiences and sensations during the use of the consumer product. In the future IT and creative designers will also be part of this team approach by providing apps via QR codes on the packaging and similar information-related contributions,
   - **Assisting in product development?**
     The modern world of product development falls apart in several different process streams. The so called fuzzy front end often allows a maximum of influence factors and creativity, while the development process itself is more and more shifting to a highly focused and targeted execution of the activity stream with limited creativity levers because of the costs connected to it.
   - **Assisting in consumer and market research?**
     Consumer and market research in the food industry requires in many cases transfer of findings from the virtual world of data into the real world of perception.

Designers are working in both worlds, the virtual, and the real product-related world. They are able to create examples and samples, which can be tested and experienced by consumers way before the final product launch.

- **Consulting to marketing?**
  Marketing consultation can be considered at a level where the validation of marketing concepts plays an important role. For this purpose modern design principles can be applied.

- **Co-decision maker at board level?**
  Depending on the positioning of a product, the role as a decision-maker at board level can be considered more or less meaningful and valuable. For new product introductions, the holistic view is mandatory in order to avoid a mismatch of product properties, positioning, and consumer perception.

- **Being a member of the executive board?**
  In some cases designers can contribute a visionary element even at board level. The interaction and team-work with commercial and technical representatives is essential for a successful execution of joint goals.

- **Anything else?**
  From my perspective designers are messengers and translators, who are helping to transport the intrinsic properties of new products like new taste or technology into the visual or haptic world for a faster and better product understanding by the customer.

- **All of the above?**
  No further comment
  The "other voices":
- *Yes to all questions.*
- Yes to the first four questions; No to questions: co-decision-maker at board level and being a member of an executive board.
  Anything else: Disrupters, push and challenge our perceptions of what is possible.
- *In my humble opinion, yes to all of the above questions. The real question is not what designers can do, but what they are allowed to do. Most companies would rely on outside consulting and would lack the understanding and/or the ability to fully incorporate and benefit from this vast and important resource.*
- All of the above. Since I took over from my predecessor, I have defined three roles for industrial designers in my present company. **(Note from the author: see also fifth block of comments under question 10)**:
  - *Design strategists* come from a background in design and usually have a master's degree or significant experience within a large company, running design and innovation projects. They employ and nurture a design-thinking mindset, not only in what they do, but also in all that everyone does within the organization. Their role is to facilitate interdisciplinary thinking and discussions and help teams navigate through the ambiguity of early projects and innovation initiatives. They analyze markets, user-needs, and internal capabilities to deliver upon new opportunity areas.
  - *Design creatives* come from a design background in either industrial or product design. They have an inherent curiosity for how things work and are trained to make things work. These designers are responsible for working closely and collaboratively with people in order to build a strong sense of empathy for those

they work with and for, and ultimately designing a new experience or product that is better than the one before … They work in a tangible world and are responsible for developing thoughts into tangible and functional form.

- *Design managers* are responsible for design teams, they act as a filter for the demands, manage resources/allocate people to projects, manage budget, promote design , network, and challenge demand.
- *First answer: all of the above (but unrealistic).*
  Second answer: designers should be working more on designing business models, novel routes to markets, and should create the right conversation language with the consumer, not designing corporate identities.

6. **How do you personally support and encourage innovation in your role or function? Can you give one or two specific examples?**
   - The flavor and fragrance industry has a long tradition of bringing innovation to the food and beverage industry. The introduction of vanillin as the first synthetic flavoring substance stimulated the development of so-called vanilla sugar ("Vanillezucker") at Dr. Oetker in Bielefeld. The combination of green color and a lemon flavor in a beer application, like in Becks Green Lemon boosted the sales of beer shandies. Both product innovations came from my flavor company at very different development stages of the flavor industry.
   - *Supported establishment of innovation centers in three locations in EMEA (Europe, Middle East) region. Personally engage with customers in discussions on innovation.*
   - Encourage my supply chain colleagues to build a culture of innovation into their day-to-day thinking and work. For example, at each operational review ask: what can we do differently next time to improve? Encourage supply chain to contribute to the R&D and design early enough to have a positive influence on the supply chain impacts of particular products.
   - *Yes, I do. Most recently I worked on changing the mindset of the top management in order to be able to start new concepts/product lines beyond the core business. In addition, strategic assessment of several technologies, processes, and new products.*
   - As Global Head of Industrial Design, my role is to help our businesses develop their pipeline of innovations, as they are very often struggling to activate the opportunity areas identified out of their consumer research. In this capacity, I am helping SBUs/Markets and Tech teams/PTCs enter into a proper dialog to engage in innovation, defining the approach, the best tools and methodologies, setting priorities, understanding each other, acting as the connector between the user need and the technological capabilities, challenging both communities to reach a common ground, and move on to launch on the market. The challenge is very often to understand the technical setting and remain the guardian of the original design intent as expressed from the consumer insights learning.
   - *After my retirement, I am coaching and consulting on innovation processes and idea-carpentry.*

7. **Do you take a personal interest in supporting and coaching innovators in your work environment?**
   - Mentoring and coaching is essential for building and maintaining a successful team. My personal experience is that individuals come with an extremely different set of capabilities and professional background when it comes to the innovator's

job, which is described in the Latin word "*innovare,*" which means renew concepts, products, and finally thinking. The first step, which I take in all mentoring tasks, is to explain and visualize the difference between their own perception of things and the perceptions of other people and consumers. In the next step I try to train precision in personal observation and professionalism in measuring correlations as is practiced in science. In the last step I focus on an in-depth understanding of the intrinsic nature of ideas and inventions versus projects and processes.

- *Yes.*
- Yes, one of my roles is to work in the supply chain to encourage a culture of innovation.
- *Indeed. It takes a lot of coaching, and it can be quite frustrating. Most people are working on innovation without proper know-how and/or formal education. Moreover, most top managers have a title that includes "innovation," with no adequate training and exposure.*
- Of course, an important part of my role is to develop best practices and to train, educate, and coach our designers in interacting with these.
- *Yes, as far as it is part of my consulting projects*

8. **How would you define the optimal and most appropriate way of compensating/coaching and mentoring innovators, be it individuals or teams? Can you give one or two specific suggestions, which, in your experience have worked or can work?**

- For the compensation of innovators I would like to suggest different things for the main activity areas. The first cluster deals with ideas and inventions, which in many cases lead to patents or so called IP. In this case compensation can be quantified by the commercial value of the final execution of the described invention. The second cluster refers to projects with the focus on renewing existing products and concepts. In this case a certain level of incentive for the team and also individuals can be considered. The whole area of mentoring, coaching, and teaching of talents is characterized by a personal relationship between the mentor and the candidate. The content develops over time and in many cases it can be extremely time consuming. Therefore a time- and quality-related compensation in many cases shows the best results and usually finds mutual understanding and wide acceptance.
- *Give space to innovate, physical and virtual, but ensure there is a reporting link and milestones. Recognition is not always related to monetary reward, it is more effective to be recognized by peers internally and externally. For example, the "Green Chemistry Award" in the USA, the "Queen's Award for Innovation" in the UK.*
- I would encourage end-to-end thinking about the design process—not just about the product in the consumer's hands but how the product works—and understanding of the impacts of design choices at each stage of the supply chain.
- *I believe that there are several ways: 1. by example starting from the concepts, brainstorming, consumer research, production, and marketing; 2. specific in-house courses; 3. reaching out and participating in an open-innovation project; 4. spending time at a design company; 5. participation in an innovation network; 6. participating in professional conventions in other fields.*
- The core skills of innovators are passion and self-motivation. I am a firm believer that the best reward for any innovator is to see his product on shelf, or his service

in operation. I don't think money or pride along the way can be any better than to see the real output of the business with consumers. That's what any innovator should be obsessed with. What works is a small cross-functional team, focused on this objective, with all relevant competences being represented, and dedicated to a particular project.

- *One of the most gratifying ways to compensate innovators is to let them build an integral process from their ideas to the market launches. Responsibility and direct feedback are key drivers of any successful innovation. Like in the case of an artist who builds his/her art from idea to finished work. The organizational structures of large corporations are genuinely poisonous for creativity and innovation because they segment the whole process for command and control reasons.*

9. **How much time do you personally spend in coaching and mentoring talent?**
   - I am responsible for the training of young trainees in my current role, which is a team effort. At the same time I coach young scientists in their roles as a group leader.
   - *30% of my time.*
   - One hour per week.
   - *Depends on the occasion/opportunity. In the food industry, this is quite limited as everyone is busy doing the "right" things. Much more time available at the university and innovation networks.*
   - Probably 30% of my time.
   - *N/A (depends on the nature of the consulting contract).*

10. **How do you see the role of design and designers evolving in the FMCG industry, and, if knowledgeable, especially in the food industry?**
    - My personal experience is that in many cases designers position themselves as a type of exotic, out-of-the-box thinking individual, bringing "color to the grey of the day." I think that this is a logical consequence of the job, which is expected. At the same time I have seen very professional and technically experienced designers with capabilities to step in and take a leadership role. There is a strong trend in the food industry to move to sustainable sourcing and production. This results in a massive need for innovation of products and concepts, which could come from a new generation of designers, who try to capture the intrinsic good side of products and stay away from manipulating consumer perception. NGOs and consumer groups are getting more and more sensitive about this point and expect transparency, as well as authenticity.
    - *Design is already an important part of these industries. The need for differentiation will drive innovation in every area, including design.*
    - No comment.
    - *This is a crucial aspect that the food industry needs to embrace. I strongly believe that it will come with real hands-on experiences. We need a new generation of designers who understand the food industry and can naturally and completely interface with business.*
    - See point 5; (**Note from the author: fifth block of comments under question 5 above**) moreover, I effectively see that the world of design has significantly evolved with integration of design thinking, which opens new roles and positions for anybody with a design mindset, whether he is a designer or not. This means we will see more design strategists positions appear in the near future, and structures will

organize around this to better integrate design within the business. This will result in a better integration of designers at all levels of the conversation on innovation, and will lead ultimately to senior design leaders sitting on the board. Having said that, I see the food industry definitely being more a follower than a leader in that field, but this industry will be inspired by the likes of Apple, BWM, and many others, where design plays a more prominent role in the business strategy.

- *I only can hope that the voices of designers are heard in the gobbledygook of modern innovation consulting advice. It should be easy for the food industry, because it has design at its heart, stemming from its origin in culinary arts.*

Index

References to figures are given in italic type.

3D printing, 29–30, 253
Adams, Edward A. 'Tink', 66
Adria, Ferran, 39
American Express, 243
Ancient Agora Museum, 43
Ansari XPRIZE, 27
APIfork, 254
Apollo missions, 145
Apple, 9, 31, 44–5, 125–6
    "Think Different" campaign, 32
architects, 24
ARK, The, 250
ARkStorm, 75
Art Center, 7, 29, 66–8
   Design-based learning lab, 84
   Nestlé and, 71–2
   transformative projects, 74–5
AT&T, 243
Autodesk 123D, 254
automotive industry, 2–3
awards, 284

baby formula, 147
bad goods, 101
baking, 166
BaldeMovil, 73–4
Ball, Roger, 26
bar code, 167
barrels, 166
Bayer AG, 2–3
Beba HA, 147
Becel/Flora, 276–7
Behar, Yves, 25–6
Belcampo Meat Co., 249

Berners-Lee, Tim, 114
Bespoke Innovations, 30
Beyond Eggs, 251
Beyond Pretty, 77
Bloomberg Business Week, 241
box metaphor, 214–15
Boyd, Drew, 213
BPA (bisphenol-A), 51
brainstorming, 205–6
   Fastpack, 208–9
   IdeaStore, 209–13
Branson, Richard, 27
Braun, 31
breakage, 100
British Airways, 243
Brown, Tim, 78
Buchman, Lorne M., 82
business plans, 143–5

California College of Art, 83
canning, 166, 167
carbonated water, 167
Carlson, Curtis, 206–7
Cavaletto, Mike, 274
Cereal Partners Worldwide, 141–2
Chefler, 253
C.H.E.F.S. principles, 279–80
chief design officer (CDO), 271
chief executive officer (CEO), 117–18
chocolate chips, 167
Churchill, Winston, 19
Citi, 243
Clarke, Arthur C., 281
Clarkson, Jeremy, 157

Food Industry Design, Technology and Innovation, First Edition.
Helmut Traitler, Birgit Coleman and Karen Hofmann.
© 2015 John Wiley & Sons, Inc. Published 2015 by John Wiley & Sons, Inc.